The
Neuroscientist
Who
Lost Her Mind

The
Neuroscientist
Who
Lost Her Mind

MY TALE OF MADNESS
AND RECOVERY

BARBARA K. LIPSKA
with Elaine McArdle

Houghton Mifflin Harcourt
Boston   New York
2018

For information about permission to reproduce selections
from this book, write to trade.permissions@hmhco.com or to
Permissions, Houghton Mifflin Harcourt Publishing Company,
3 Park Avenue, 19th Floor, New York, New York 10016.

hmhco.com

Photographs throughout the book are courtesy of the author.
The illustration on page 7 is © Witek Lipski.

Library of Congress Cataloging-in-Publication Data
Names: Lipska, Barbara K., author. | McArdle, Elaine, author.
Title: The neuroscientist who lost her mind : my tale of madness and recovery /
Barbara K. Lipska with Elaine McArdle.
Description: Boston : Houghton Mifflin Harcourt, 2018. |
Includes bibliographical references and index.
Identifiers: LCCN 2017054093 (print) | LCCN 2017046211 (ebook) |
ISBN 9781328787279 (ebook) | ISBN 9781328787309 (hardcover)
Subjects: LCSH: Lipska, Barbara K.—Health. | Melanoma—Patients—
Biography. | Brain metastasis—Patients—Biography. | Neuroscientists—
Biography. | BISAC: BIOGRAPHY & AUTOBIOGRAPHY / Personal Memoirs. |
SCIENCE / Life Sciences / Neuroscience. | PSYCHOLOGY / Mental Health. |
BIOGRAPHY & AUTOBIOGRAPHY / Medical.
Classification: LCC RC280.M37 (print) | LCC RC280.M37 L57 2018 (ebook) |
DDC 616.99/4770092 [B]—dc23
LC record available at https://lccn.loc.gov/2017054093

Book design by Chrissy Kurpeski

Printed in the United States of America
DOC 10 9 8 7 6 5 4 3 2 1

Contents

To Mirek, my rock

To science, for saving lives

In memory of Witold,
for whom scientific advances came too late

Prologue

I'*m running and running and running.* For hours, I've been running. I want to get home but I have no idea where that is, even though I've lived in this neighborhood for twenty years. So I keep running.

I'm roaming these tree-lined streets in suburban Virginia at a fast clip, wearing my usual outfit—a tank top and running shorts. I sweat as my pace increases, faster, then faster still, my heart pounding but my breath even and unhurried as I sail past large homes with two-car garages and bicycles parked in driveways.

It's the end of spring 2015 and the beginning of what will become a particularly hot and humid summer. The grass on the immaculately trimmed lawns is still green and lush. Pink and white peonies are in full bloom, and all around me azaleas explode in a rainbow of colors.

I've jogged this route hundreds of times over the past two decades. I should recognize each maple tree and camellia bush on each street corner, and every gash in a curb where a teenage driver took a corner too fast. They should be landmarks as familiar to me as anything in my life. But today it's as if I've never seen them before.

When my husband and I bought our home here twenty-five years ago, just two years after leaving the grimness of Communist Poland, this normal American suburb seemed a dream come true. What luxuries it contained! Settled into our new home, we quickly adopted a middle-class American lifestyle, complete with regular meals of Chinese takeout and buckets of ice cream—indulgences that were nonexistent in Eastern Europe.

One day, I saw a photo of myself—arms chubby and dimpled, thighs spread across my chair—and was shocked into a major lifestyle change. I needed to get more exercise, and I began to run. Not one for minor shifts in my life, I decided I would enter a race as soon as I was able.

At first, I couldn't jog a single block. Within a year, I was running three miles. After two years, I signed up for my first race, a six-mile competition where I finished at the top of my age group. Since then, my entire family has become dedicated athletes. Runners, cyclists, and swimmers, we're always training for one competition or another.

And so, each morning, I run.

A creature of routine, I always start by taking my German-made prosthetic breast from the shelf in my bathroom. I've worn the breast ever since undergoing a mastectomy following a battle with breast cancer in 2009. Fashioned from high-tech plastic, it is flesh-colored and feels like a real breast, and it is proportioned to match the breast on my right. It even has a tiny nipple. Engineered for athletes, it's light and has a special adhesive on the underside to hold it on to my body. Every morning before my jog, I slap it into place on the smooth, flat skin of my left chest before donning my clothes and sneakers. And then I'm off.

But this morning—this morning—began differently.

After pouring my usual glass of water, I headed into the bathroom and peered at myself in the mirror.

My roots are showing, I thought. *I need to dye my hair.*

Now!

I mixed the dye—a brand of henna from Whole Foods that gives my hair a funny purple tint that I love—in a small plastic cup, then squirted it onto my scalp and spread it over my head. I pulled a plastic bag over my skull and tied it with a little knot on one side to hold it in place.

I must hurry. It's urgent—urgent!—to get outside and begin running!

I grabbed my shirt and shorts and headed back into the bathroom.

I looked at the breast on the shelf.

No. Too much trouble. It weighs me down. I'm not going to spend precious time on stupid things like that.

I quickly pulled my tight-fitting shirt over the plastic bag on my head. My body was noticeably lopsided without the prosthetic breast, but I didn't think twice about it.

I need to leave now!

Purple-red dye oozed down my face and neck as I sprinted out of the house and down the street.

Now, as I run along in the morning heat, the dye spreads over my shirt and stains my asymmetrical chest.

The streets are almost empty in our sleepy neighborhood. If any of the few people I do pass are surprised by my strange appearance, I don't notice. I glide along, absorbed in my own internal world.

After an hour I begin to tire and I am ready to return home. But my neighborhood looks strange. I don't recognize these streets. I don't recognize these houses.

I have no idea where I am. So I keep moving.

It's preposterous that I could get lost in this familiar place, but that fact barely registers in my mind. With no plan for where I'm headed, I simply continue to run.

For another hour or more, I jog along, misshapen and covered in gore. I'm oblivious, unaware of anything amiss. I just run and run, my thoughts drifting into open spaces and big skies.

Somehow, I finally come upon our two-story Colonial. I open the door and find myself in the cool, dark hallway. Tired and sweaty, I take off my sneakers and socks, which are completely soaked.

On my way upstairs, I catch a glimpse of myself in a mirror. My head is caked in sweat mixed with hair color, the plastic bag plastered on top like a weird swimming cap. Streaks of purple dye, long since dried to black, have crusted in thin rivulets down my neck and upper arms and all over my shirt, accentuating the sunken left side of my chest. My face is deep red from exertion.

Nothing strikes me as unusual. I continue past the mirror up the stairs.

In his home office, my husband, Mirek, is sitting at his computer with his back to the door. When he hears me enter the room, he says, "You've been away a long time. Good run?"

Then he turns to me with a smile—and freezes.

"What happened?" he exclaims.

"What do you mean?" I say. "It was a long run."

"Did anybody see you like this?" He seems shaken.

"Why would I care if someone saw me? What are you talking about?"

"Wash it off," he says. "Please."

"Calm down, Mirek! What are you going on about?" But I head into the bathroom to do as he asks.

What's wrong with him? Why is he acting so strange?

I emerge from the shower clean and relaxed. But something nags at me.

The man I love is alarmed. But why?

Mirek's behavior should be a red flag, a clue that something is terribly wrong. But a moment later, the unpleasant thought simply slips through the cracks of my broken mind and is gone.

I am a neuroscientist. For my entire career, I have studied mental illness, first in my homeland of Poland and then, since 1989, in the United States, at the National Institute of Mental Health (NIMH), a division of the National Institutes of Health (NIH) in Bethesda, Maryland. My specialty is schizophrenia, a devastating disease whose victims often have difficulty discerning what is real and what is not.

In June 2015, without warning, my own mind took a strange and frightening turn. As a result of metastatic melanoma in my brain, I began a descent into mental illness that lasted about two months, a bizarre tailspin that I couldn't recognize at the time. I emerged from that dark place through a combination of luck, groundbreaking scientific advances, and the vigilance and support of my family.

I'm a rare case; I lived through a terrifying dive into brain cancer and mental illness and emerged on the other side able to describe what had happened to me. According to psychiatrists and neurologists—medical doctors who specialize in the brain and nervous system—it's highly unusual for someone with such serious brain malfunction to be successfully treated and return from the shadowy world of mental impairment. Most people with as many brain tumors as I had and the serious deficits they caused simply don't get better.

As frightening as my breakdown was, I regard it as a priceless gift for a neuroscientist. I studied the brain for decades and conducted research in mental illness, but my brush with madness gave me firsthand experience of what it's actually like to lose your mind and then recover it.

Every year, approximately one in five adults worldwide experiences a mental illness, from depression to anxiety disorders, from schizophrenia to bipolar disorder. In the United States, mental illness affects nearly forty-four million adults each year, and that number does not include people with substance-abuse disorders. In Europe, 27 percent of adults experience a serious mental disorder in any given year. Mental illness often emerges during young adulthood and lasts for someone's entire life, causing tremendous suffering for the person who is ill as well as for his or her loved ones. A significant number of homeless and incarcerated people suffer from mental illness, and the societal consequences do not end there. Mental illness costs the global economy $1 trillion each year—$193.2 billion in the United States—as people who would otherwise be productive are unable to function because of their disabilities. More than just incapacitating, mental illness can also be deadly. Of the roughly 800,000 people worldwide who die each year by suicide—41,000 in the United States alone—90 percent suffer from mental illness.

The United States spends significantly more on treating mental disorders than it does on any other medical condition—a whopping $201 billion in 2013. (Heart conditions, for which the U.S. spent $147 billion that year, rank a distant second.) But even with these resources and the tremendous efforts of dedicated scientists and physicians, mental illness remains deeply enigmatic, its causes generally unknown, its cures undiscovered. Despite the overwhelming body of research on mental illness to which new findings are added almost every day, we scientists still don't understand what happens in the brains of mentally ill people. We don't really know yet which brain regions and connections are malformed or undeveloped or why the brain goes awry. Are people who become mentally ill destined to suffer because of some genetic predisposition, or did they experience something that broke their brains, mangled their neuronal connections, and altered their neurologic function?

Today, the data suggest that mental illness is caused by a combination of heredity and environment, the latter involving multiple factors—including drug use and abuse—that act in complex interplay with one another and with our genes. But it remains exceedingly hard to pinpoint the biological and chemical processes for mental illness, in part because these disorders are diagnosed through observations of behaviors rather than through more precise tests. Unlike cancer and heart disease, mental illness has no objective measures —no biological markers that we can see on imaging scans or determine through laboratory tests—to tell us who is affected and who is not. In the aggregate, groups of people who suffer from mental illness may show differences in their brain structures or functions, but for now, individual patients can't be diagnosed using conventional measures such as blood tests, computerized tomography (CT) scans, or magnetic resonance imaging (MRIs).

Diagnosing mental illness is all the more difficult because the constellation of symptoms not only varies from person to person but also often fluctuates over time within an individual. Not everyone afflicted with schizophrenia screams in distress, for instance; some people with the disease may shut down and stop communicating. Likewise, people with dementia may be attentive and engaged one minute and detached and withdrawn the next. Even more challenging, some indications of mental illness may be exaggerations of normal personality traits, making the behaviors particularly hard to recognize as pathological. With people who are naturally frank and outspoken, the lack of judgment that can accompany dementia may at first be construed as their typical bluntness. Similarly, when introverted people become more withdrawn, others may not realize that they are exhibiting symptoms of Alzheimer's disease.

For researchers, it's becoming clearer that specific mental disorders are not well-defined categories of illness, each delineated by a distinct set of symptoms and biological indicators. The same symptoms may not even be caused by the same illness, so two people who exhibit the same erratic behavior may in fact be suffering from two completely different disorders. Or perhaps there is overlap among various mental disorders in terms of symptoms, biological mechanisms, and causes. Some genetic and clinical analyses find similarities across a wide variety of diagnoses, suggesting that mental illnesses share a common neurobiological substrate. Modern science is currently exploring this possibility.

Today, scientists are quite confident that the main site of disruption in people with mental illness is the highly evolved prefrontal cortex, which sits at the front of the brain, and its network of connections with other parts of the brain. But what these abnormalities are and how exactly the brain malfunctions in various mental problems remains a puzzle.

When a person's behavioral changes are triggered by brain tumors, as mine were, it may seem easy to establish a cause-and-effect relationship between neurological factors and behaviors. Neurologists like to try to localize every problem to a particular part of the brain, and sometimes that's more or less possible. But metastatic brain tumors—whether from melanoma or breast cancer or lung cancer—tend to involve multiple parts of the brain at the same time. When you have two or more tumors, as I did, it becomes especially difficult to figure out what part of the brain is affecting what behaviors. In addition, when there is extensive swelling from tumors and treatments, the entire brain contributes to the altered behavior.

While we don't know exactly what took place in my brain or where precisely it happened, my journey has given me an invaluable opportunity to tour the landscape of the brain. As a result, I've come to better understand this breathtakingly complex structure and its incredibly resilient product: the human mind.

As with everyone who suffers from mental disorders, I experienced a constellation of symptoms during my brush with madness that were unique to my case. But during my brief mental breakdown, I exhibited many symptoms described in the *Diagnostic and Statis-*

tical Manual of Mental Disorders, fifth edition (*DSM*-5), the official guide clinicians and researchers use to classify various mental illnesses. For that reason, the similarity between my experience and that of people with a wide range of mental illnesses—from Alzheimer's to other dementias, from bipolar disorder to schizophrenia—is remarkable. Identifying these parallels and using them to better understand the experience and causes of mental illness is one of my main goals in this book.

I gained a deep understanding of what it is like to live in a world that makes no sense, that's bewildering and unfamiliar. I know what it's like to be so confused that you trust no one, least of all those closest to you, who you may be convinced are conspiring against you. I know how it feels to lose not only the powers of insight, judgment, and spatial recognition but also the faculties essential for communication, such as the ability to read. Perhaps most chilling, I also know what it's like to be completely unaware of these deficits. It was only after my mind began to return that I learned how warped my reality had been.

After I emerged from that dark space and was given a second chance at sanity, I wanted to explore, as a neuroscientist, what went wrong in my brain. I learned that my frontal and parietal lobes—which are responsible for many of our most human behaviors—were malfunctioning. This helps to explain why I behaved in ways similar to people with mental illness: why I got lost in familiar places, forgot things that had just happened to me, and became angry, mean, and unloving to my family; why I became obsessed with strange little details like what I was having for breakfast while ignoring the fact I was about to die; and, most striking, why I failed to notice any of these insidious changes in myself. Even as my mind was deteriorating, I couldn't see that I was slipping into mental illness.

In addition to providing insight into mental illnesses such as schizophrenia and dementia, my experience gave me a greater understanding of other brain disabilities, including the mental declines that many of us encounter as we age. Many people may someday face in themselves, their partners, or their parents the bewildering mental changes I had—memory loss, disinhibited and inappropriate behaviors, altered personality, and the inability to recognize these problems in oneself. The frontal cortex, the part of my brain that was most affected by my tumors and treatment-induced swelling, is one of the regions that often begin to fail as we enter our senior years (the hippocampus is another). It's one of the many ironies of my story that if I live long enough to see old age, there's a good chance I will experience many of the same mental changes all over again.

In the course of losing and regaining my sanity, I've come to identify with other people who have known mental illnesses firsthand. This sense of connection with others who suffer has spurred me to share my story. While more attention is being given to mental illness than ever before, it nonetheless continues to be stigmatized by society. Even though mental disorders are physiological in nature—they are diseases of the brain, just as coronary problems are diseases of the heart—the mentally ill are often treated as if *they* are to blame, as if *they* have done something wrong. Their families are frequently stigmatized as well. If nothing else, I hope my experience helps others recognize that mental illness is no more the patient's fault than cancer is and that the best response to mental illness is empathy and a greater commitment to finding cures.

After losing my mind and regaining it, I like to think I am more attuned to other people's feelings and troubles, that I am more understanding as a mother, wife, friend—and scientist. While I believe that I've always been compassionate toward people with mental illness, since my own brush with madness, the quality of my compassion has deepened. I also live my life more consciously, aware of how lucky I am to be reunited with my family and able to continue my life's work.

This book is an account of what mental illness looks like from the inside. But it is also a map of my evolution as a scientist and a person. It is the story of an incredible journey, one from which I could not have imagined I would ever return. It is a story that I never thought I would be able to tell, of how I went from being a scientist studying mental disorders to being a mental patient myself—and how, remarkably, I came back.

The Rat's Revenge

I sit among a thousand brains, a thousand brains of the mentally ill.

As director of the Human Brain Collection Core at the National Institute of Mental Health, I work surrounded by brains; a library of brains, a bank of brains, a compendium of brains that for any number of reasons hadn't worked the way they should have. Brains that saw hallucinations, heard mysterious voices, were buffeted by wild mood swings, or were deeply depressed. Brains that have been collected, cataloged, and stored here for the past thirty years.

About a third of these brains come from suicides. That desperate and heartbreaking act is the ultimate cost for so many people who suffer from mental illness, and my colleagues and I are reminded of this grim fact each and every day.

Each specimen arrives to us fresh and bloody, glistening inside a clear plastic bag placed carefully inside a cooler of ice. It looks like a piece of red meat, unconnected to any real humanness. Yet just a day earlier, it had directed every movement and thought of the person from whom it came.

To understand mental illness—and to treat and one day cure it— researchers need a steady supply of brains. This is where institutions like the NIMH, the leading federal agency in the United States for research on mental health, come in. At the brain bank, we gather these incredible organs, slice them into usable tissue samples, and share them with scientists around the world.

But collecting brains isn't easy. It's especially difficult to get brains that come from people with schizophrenia, bipolar disorder, major depression, anxiety disorders, and addictions to various substances—cocaine, opioids, alcohol, and even cannabis—that attract abuse. What's more, we can't use brains of mentally ill people who died of serious illnesses, who were in hospitals on ventilators, or who were heavily medicated before taking their last breath. Brains marked by other illnesses or medical issues would only add complexity to the already overwhelming puzzle that we are trying to solve: What causes mental disorders?

In order to begin to understand this, we also need brains from people without mental illness (control brains), so we can examine and compare them with diseased brains. In short, we need clean and healthy brains both with and without the terrible presence of madness.

We get most of our brains from the morgues in nearby medical examiners' offices, where bodies typically arrive because people have died under suspicious or mysterious circumstances. And so, in addition to receiving the brains of suicides, we are also the unintentional beneficiaries of homicides and unexplained deaths.

First thing each morning, the technicians in our brain bank telephone local medical examiners' offices and ask, Do you have any brains for us today?

We're in a rush. If a person has been dead more than three days, we can't use the brain. We need the brains before the tissue begins to decompose, before their proteins and other molecules, the ribonucleic acid (RNA) and deoxyribonucleic acid (DNA), begin to break down, rendering them useless for molecular studies.

The morgue workers tell the techs about the bodies that have arrived in the past twenty-four hours, sharing what information they have. Often, it isn't much, just the barest of facts: A young man who overdosed on heroin. A middle-aged woman with a heart attack. A teen who hanged herself. At this point, it may be all we know about each person.

Once the technicians have compiled their list of candidates, they come to me, and together we narrow it down. Do we want this one, a drug overdose? Or this one, an elderly man whose wife told morgue investigators he was an alcoholic? Here's a man who died in a car accident. There's no indication he had mental illness, so maybe researchers will be able to use his brain as a control in their studies. But he might have sustained a head injury; do we still want him?

If there's any possibility that a brain may be right for us, I usually say yes. The brains we seek are rare and precious, and we don't get nearly enough.

Once we have settled on potential candidates, our technicians contact each one's next of kin to make a wrenching request: Would you consider donating your loved one's brain to medical research?

It seems a simple question. Yet a few hours earlier, these people were alive. Now they are forever lost, and we are asking parents or spouses or children to see through their own shock and grief to give us the most personal part of their loved ones, the part that made up the very essence of who they were. Not surprisingly, perhaps, only about a third of them agree to donate the brains we seek.

When a brain arrives at our bank, we label it with a number in order to protect confidentiality. Then our job begins in earnest. We can now cut this specimen open and study its inner workings in an attempt to better understand mental illness.

It is among these brains—sliced up and frozen in a slurry of hope and optimism that they will one day reveal their secrets—that I do my work.

Brains are a bloody business. I've worked with them for over thirty years, starting with rat brains, each of which is the size of a walnut, smooth, and relatively simple. They have none of the intricate folds and crevices—called gyri and sulci—of the human brain.

By contrast, the human brain is large, elaborate, and far more complex. It is a feat of evolutionary engineering. All of its folds, all of those gyri and sulci, ridges and crevices, help to squeeze more storage and function into the relatively small space of the human skull. Consciousness is one of the many products of this marvelously complicated piece of tissue. Unfortunately, mental illness—an affliction of consciousness—is a product as well.

In our quest to understand what's wrong in the brains of people with mental illness, we have to dig deep into the brain's tissues, cells, and molecules. Novel techniques make this a little easier every year. To try to unlock the secrets of schizophrenia, for instance, I examine thin slices of brain stained with radioactive or fluorescent dyes and evaluate the cells' various molecules, proteins, and types of RNA and DNA. To read their genetic code, I analyze the brain cells' minute molecular composition with modern sequencing machines.

As a neuroscientist and molecular biologist, I'm an expert on the brain. But I'm not a clinical doctor. Before I became head of the brain bank, I'd never worked with intact human bodies or even identifiable body parts. I did my work in quiet laboratories far from morgues and hospitals, and by the time the brains got to me, they didn't look like brains at all. They were pulverized bits of frozen tissue that looked like specks of pinkish flour suspended in liquid in little test tubes, or they were thin slices of tissue preserved in foul-smelling chemicals. They could have been almost anything or come from almost any organism.

It never bothered me to be both intimate with and distant from the subjects of my studies. After all, that is the nature of scientific research. Each scientist works on her own small, discrete piece of an overwhelming puzzle that she hopes will someday be solved by researchers' collective efforts and to which her narrow contribution will have been some significant part.

Before I took this job, I'd never even touched a whole human brain. I'd been to a morgue several times, seen bodies splayed open with their organs removed. But I'd never seen a brain lifted out of a skull. I'd never held a whole brain in my hands, much less cut one apart.

"You have to do it yourself," my predecessor at the brain bank, Dr. Mary Herman Rubinstein (known as Dr. Herman), urged me in 2013 as she trained me. "When we get the next brain, we'll slice it up and freeze it together."

So we do. On a sunny day in September of that year, with the leaves just beginning to turn yellow and red but the air still warm and comforting, we stand in the lab awaiting the arrival of my first brain. We are armored in protective gear—surgical masks strung from ear to ear, plastic shields over our faces, hair caps secured tightly around our foreheads, several layers of latex gloves that cover our arms up to our elbows, white lab coats overlaid with plastic aprons to protect us from splashing blood, and plastic booties covering our shoes.

A technician carries in a large white cooler, the kind that holds beer and steaks for a football tailgate party. This cooler, I know, contains a human brain packed in lots and lots of ice.

It is critical that the brain stays cold, because this helps slow the process by which tissues break down. For our experiments, the brain cells' RNA—key to how genes are expressed—must be intact. Putting a brain on ice immediately after it's removed from the body is the first step in preserving the RNA, but for long-term storage, we must quickly deep-freeze the tissue. Keeping the brain at very low temperatures can halt RNA decomposition for decades.

Dr. Herman opens the lid of the cooler and carefully lifts out a clear plastic bag frosted with ice. She slowly takes out the brain and places it in my outstretched palms. It fits comfortably in my hands. Heavy, cold, and wet, it drips with blood just like any other piece of meat. The average brain weighs 1,300 grams, or about three pounds; in time, I will see some that are as large as 1,800 grams, about four pounds. It feels like firm Jell-O, but in fact it's quite fragile; if I'm not careful, parts might snap off.

Given that the human brain is the most complex structure in our universe, you'd expect it to look more . . . well, complicated. But it just doesn't appear all that extraordinary. The first time I saw a dead body in a morgue—all the blood, muscles, bones, and skin—I was afraid I would faint. But I find the brain now in my hands much less disturbing. Removed from the body in which it grew, this brain seems almost nonhuman.

Yet the enormous contrast between this ordinary-looking piece of meat and the complexity within it is deeply moving. It is awe-inspiring, marvelous, to realize that everything about a human being can be contained in my hands.

This brain governed a person who was alive less than a day ago. That much I'm sure of. But what else can I know about the brain that I hold? Did it come from a woman or a man? Did this person suffer from mental illness? Did this person kill him- or herself? The likelihood of that is high, given where we get these brains. But it's also entirely possible that the brain came from an elderly woman who died of pneumonia or a young man who was killed by a gunshot wound to his chest. The person might have suffered from schizophrenia or depression, but he also might have had a clean bill of mental health. There is no way to know from looking at it with the naked eye. The brain does not reveal its secrets easily.

A whole brain is shaped somewhat like a football and is divided by a deep groove down the middle into a left and right hemisphere. Each hemisphere has four lobes: the frontal, temporal, parietal, and occipital.

As I hold this brain in my hands, I stare at the frontal lobes, the largest of the lobes. These regions of the cerebral cortex, the outer covering of the brain, determine much of our species' conscious existence, from our perceptions of the world to our most private thoughts and imaginings. They are the parts that fascinate me the most and that preoccupy the overwhelming majority of neuroscientists.

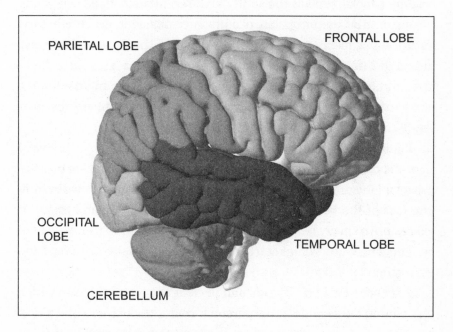

The major regions of the human brain.

The frontal lobes—one on the left, one on the right—extend from the bottom of the forehead, right above the eyes, all the way back to the top of the skull. Like the other lobes, they're wrapped around the more primitive parts located farther inside the brain.

I linger over the frontal cortex, the front top part of the frontal lobe, situated roughly where the hairline is. Large and full of folds and crevices, it is both the youngest and the most evolved part of the human brain. It determines who humans are—thinking, remembering, problem-solving creatures capable of judgment and informed decision-making.

The prefrontal cortex, the foremost part of the frontal cortex, sits just behind the forehead. This relatively small part of the cortex is perhaps the most crucial to our sanity because the prefrontal cortex controls what is known as executive function—the most complex cognitive tasks, such as the ability to differentiate between right and wrong, inhibit inappropriate or impulsive behavior, and predict the future consequences of things that happen in the present. Extensive research into the neuroscience of mental illness leaves little doubt that problems of the prefrontal cortex are central to mental illness. But we don't know what kinds of problems they are, and just by looking at this brain's frontal cortex, I certainly can't tell.

Behind the frontal lobe, separated by a deep sulcus, or groove, I spot the parietal lobe, another large chunk of convoluted cortex. The parietal lobe coordinates sensory information sent to the brain from the rest of the body, allowing us to feel, taste, move, and touch. It places us in space, tells us where we are in relation to things around us, and where our bodies start and end. It also enables us to read and do math.

I turn the brain on its side and peek at the temporal lobe, which lies behind the temple, roughly above where the ear is. This part of the cortex is responsible for high-level auditory processing, for hearing and understanding speech. Beneath it, deep inside the brain, hidden from my eyes and surrounded by layers of cortical tissue, sits the hippocampus, named from the Greek word for "seahorse" because of its unusual curved shape. An evolutionarily primitive part of the brain, the hippocampus stores long-term memories. It also works like a GPS, enabling spatial navigation so we know where we are.

Hidden at the back of the brain is the exquisitely ribbed cerebellum, made of densely packed neurons. It coordinates voluntary movements: how we sit, walk, and speak. Just above it, where one would tie a ponytail, is the fourth and final lobe, the occipital lobe, the structure that processes information from the eyes and enables us to see.

All of the brain structures are enormously important to everyday functioning. If you injure the brain stem at the back of your brain—the part that regulates breathing, heart rate, and other basic functions—you could be paralyzed, or die. But the frontal cortex is perhaps the most precious brain region of all. While a person won't die without a frontal cortex, it is the part that makes us human. Damage to this region of the brain results in a large number of adverse symptoms, from memory loss to the inability to plan and organize actions, from problems with language and speaking to inappropriate behavior and poor judgment.

I would be happy to linger longer in admiration of this brain, the first I've ever held, but Dr. Herman and I must work quickly to preserve the specimens for our studies.

I carefully place the brain on a large board that sits atop a bed of ice and pick up the dissecting knife, which is very long with a razor-sharp edge.

"Pretend you're slicing bread or steak," Dr. Herman instructs me. "Keep the knife's edge perpendicular to the top surface of the brain, and try to make each cut parallel to the previous one."

Holding the brain with my left hand, I lift the knife and then begin to slice. The cold storage has made the tissue firm, and the knife slides through easily.

My first cut is longitudinal along the crevice that separates the brain's two hemispheres. I then slice into the left hemisphere, from front to back, creating uniform slices about half an inch thick. After a while, I feel the brain becoming mushy as it warms. Instead of falling neatly onto the cutting board, the slices fold over and crumple. I continue, though, getting better with each cut.

I pick up and examine each slice, and Dr. Herman points to its folds and creases and the borders marked by different tones of pinkish gray or white. These delineate subregions of the brain, the gray, neuron-rich areas and the white connecting fibers that run between them. Depending on where a given slice is, a particular specimen may include parts of the hippocampus, the amygdala, or some other structures inside the brain.

We quickly place each slice on a glass plate and immerse it in a mixture of dry ice and a volatile chemical called isopentane—a slurry with a very low temperature of minus 86°C. The semiliquid mixture steams and bubbles violently as we slide the tissue into it, and the slice freezes instantly, turning in seconds from bloody pink to frosted white. This procedure preserves the anatomy of the slice, preventing cell membranes from bursting open as they would during a slower freezing process. We promptly fish it out with forceps and place it in a plastic bag that we seal shut and label with a printed barcode. The preservation process is now complete.

If this brain initially resembled an ordinary piece of meat, it now looks like a stack of cold cuts in a grocery store's deli case. As if to reinforce that impression, white-coated technicians arrive to ferry the cut-up samples to our laboratory's deep freezer. There the specimens will sit, cold and silent, until they can be employed in our endless quest to discover the brain's secrets.

Human brains are exquisitely complex, but we can learn a great deal about them by studying creatures with brains vastly simpler than our own, as I found out early in my scientific career.

Thirty years before I became the leader of the brain bank at NIMH, I was a young research scientist at the Institute of Psychiatry and Neurology in Warsaw with a master's degree in chemistry and a PhD in medical sciences with a focus on the brain and nervous system. In the mid-1980s, I was working on clinical trials of drugs manufactured by Western companies to treat schizophrenia and living in a small apartment in Warsaw with Mirek, my then boyfriend, and my two young children from my first marriage.

In August of 1988, our lives were upended. That month, at the invitation of a German pharmaceutical company, I attended the International Congress of Neuropsychopharmacology in Munich. I was to present a poster on certain antipsychotic drugs designed to reduce the severity of hallucinations and psychosis, the most distressing symptoms of schizophrenia. I had no way of knowing that my focus was soon to shift from treating this terrible disease to hunting for its underlying causes.

I arrived in Munich with no more than twenty dollars in my pocket—an entire month's salary—and was immediately dazzled by the opulence of West Germany. But my culture shock paled in comparison to the thrill I experienced when, at the conference, I was approached by Dr. Daniel R. Weinberger, an NIMH psychiatrist who was world-renowned for his studies on schizophrenia. No sooner had we met than Dr. Weinberger offhandedly suggested I come work as a postdoc in his lab.

I could hardly believe my luck. NIH was the most prestigious medical institution in the world, and its mental-health division was at the forefront of global research on the very illnesses that I'd devoted my career to studying. I'd never dared to dream that I might someday end up at NIMH myself.

A few days later, I returned to Poland and proudly announced to Mirek and my children that we were going to America! They were just as excited as I was. Poland at the time was looking bleaker and more unstable than ever, and many of its unhappy citizens dreamed of the freedom that the West offered. And everyone knew that American society was the freest of them all.

I arrived in the United States ahead of my family, in the spring of 1989, just as Poland was tipping toward democracy and threatening to bring the rest of the Soviet bloc down. The day after my arrival, Dr. Weinberger—who would be my boss for the next twenty-three years—drove me to the NIH campus and introduced me to Dr. George Jaskiw, a psychiatry fellow from Canada. Dr. Jaskiw became an enthusiastic mentor to me, and together we began to explore the mysteries of the same disease—schizophrenia—that I had studied in drug trials in Warsaw.

Dr. Jaskiw and I worked on rats because their brains are similar to human brains in their structure, although not nearly as sophisticated, and because rats display complex behaviors, such as working memory, cognition, and social behaviors, that are useful in understanding humans. We first focused on creating slight defects in the hippocampi of living rats because robust research data at that time suggested that the hippocampi in humans with schizophrenia were structurally abnormal and therefore did not function correctly. To disrupt the connections between the hippocampus and the prefrontal cortex in the brains of newborn rats, we injected minute amounts of neurotoxins into the hippocampus. In this way, we created brains with faulty wiring between these two areas critical in schizophrenia. We wanted to see how different our neurologically altered rats would be from normal animals and, especially, how they would behave when they grew up.

I'd never sliced into any creature, living or dead, but I was delighted to be part of this work. We threw ourselves into the experiments with the crazed abandon of knowledge-hungry scientists. Once, when I needed a quiet area to conduct our rat-behavior experiments, I placed my rats in their testing cages on the floor of the men's restroom, taped up a sign that said EXPERIMENT IN PROGRESS! DO NOT ENTER, and locked the door. I was determined to learn and succeed. Dr. Jaskiw taught me neuroanatomy and neurochemistry, rat physiology, and the best techniques for brain dissection. Together we operated on and tested thousands of rats.

After eighteen months, and much to my dismay, Dr. Jaskiw left NIH for another career opportunity. Without him, my work became much more challenging. At times, I wept with frustration as I tried to recognize tiny structures in rodent brains, use our lab's finicky slicing machines, and catch escaped rats as they hid under the cabinets, hissing and baring their razor-sharp teeth.

As painful as Dr. Jaskiw's departure was, it forced me into independence—and led to the most significant discovery of my career. Just as we had expected, this scientific breakthrough concerned the frontal cortex, the same brain region whose critical importance I would come to understand on a deeply personal level when, ironically, my own began to break down.

Schizophrenia is a devastating illness that has plagued humans for many thousands of years. Today, it affects about 1 percent of the population worldwide—over seventy million people, including more than three million in the United States and over seven million in Europe. Schizophrenia can affect individuals in any area and from any culture or social class. Symptoms vary from person to person, as does responsiveness to treatment. Many patients suffer from delusions, hallucinations, and full-blown psychosis—the symptoms you see exhibited, for example, by people wandering the streets talking to themselves. Many patients with schizophrenia show cognitive deficits and are unable to make decisions and think logically. The deficits may particularly affect working memory, which helps prioritize and execute the tasks of life. A significant number are depressed and have trouble displaying emotions.

Until quite recently, psychiatrists believed that schizophrenia was a psychological illness caused by stress and upbringing, particularly by the influence of a "schizophrenogenic mother" who did not provide her child with enough maternal warmth and care. Today, this theory has been soundly discredited. Schizophrenia, we now know, is a disease caused by abnormal brain structure and function, just as heart disease is a product of faulty arteries. The difference is that we don't yet have a "brain fingerprint" for schizophrenia.

In the 1940s and 1950s, doctors suspected (correctly) that the frontal cortex was involved in mental illnesses, including schizophrenia. They began treating such illnesses, at times, with a frontal lobotomy—a horribly invasive type of brain surgery that involves cutting at least some of the connections within the prefrontal cortex or from the prefrontal cortex to other parts of the brain. Controversial from the start, lobotomies robbed some patients of their personalities and intellects. (These appalling effects did not stop the Swedish Academy from awarding António Egas Moniz, the neurologist who developed the procedure, a Nobel Prize in 1949.)

The advent of antipsychotic drugs in the mid-1950s, which alleviated at least some psychotic symptoms in most patients, helped sideline this crude and brutal "cure." But that pharmaceutical breakthrough came too late for many people. Between 1946 and 1956, an estimated sixty to eighty thousand lobotomies were performed worldwide.

Since the mid-1990s, the focus of research in mental illness has shifted from psychological studies, which analyze behaviors, to genetics and the study of chemicals in the brain (DNA, RNA, and proteins). This allows us to search for inherited genes, mutated genes, aberrantly structured proteins, or dysfunctional pathways that are associated with an increased risk for mental illness. The hope is that, by using precisely targeted therapies that activate or inhibit certain molecules, we can bring these disrupted pathways back to normal.

Still, for the most part, scientists' understanding of the causes of schizophrenia (as well as of other mental disorders) remains woefully inadequate. Abnormalities in perhaps hundreds or even thousands of genes may be required in order for schizophrenia to manifest itself in a particular person. And because of the great variability among the individual genetic makeups of people afflicted with schizophrenia, it is impossible at this time to predict whether any given individual carries enough risk variants to make him or her ill.

The experiments I conducted in the 1990s on rodents provided clear evidence that abnormal behaviors in rats and, by extension, in humans, may be triggered by subtle brain insults that result in lasting cognitive deficits. The rats whose brains we altered demonstrated difficulties in spatial recognition, including finding their way through mazes that contained savory rewards. Compared to normal rats, they also showed a lack of interest in novel places and objects, and they didn't interact as much with their peers. We concluded that in humans, as in our rats, slight brain defects may be initiated by a variety of factors that compromise the function of the developing brain, leaving it permanently "out of whack." Those factors in humans might include maternal malnutrition or viral infections and perhaps many other influences in combination with defective genes that alter molecular pathways and the wiring within and between brain regions. Our findings clearly implicated the frontal cortex as an essential site for the development of schizophrenia, just as Dr. Weinberger and my NIMH colleagues hypothesized in the late 1980s.

Our discoveries gained enormous interest around the world and became known as the neonatal hippocampal lesion model of schizophrenia or, for short, the Lipska model. Drs. Jaskiw and Weinberger and I first described our findings in a paper published in 1993 in *Neuropsychopharmacology*, the official publication of the American College of Neuropsychopharmacology. Since then, the Lipska model has been described in hundreds of scientific publications, replicated in many laboratories around the world, and applied to other research areas, including electrophysiology, genetics, and cognition. It has also provided a framework for designing new drugs that might offer benefits for treatment of cognitive deficits in schizophrenia. In 1996, our model was awarded a U.S. patent for screening and developing novel antipsychotic treatments.

In 2002, I became director of the molecular biology lab at the NIMH, where I continued to study chemical and genetic differences in brains of people with mental illness. The subsequent decade was a busy and fruitful one for me, despite my own serious encounters with illness: breast cancer in 2009, and melanoma—the deadliest form of skin cancer—in 2011. I was convinced I'd beaten them both, and I kept my eyes fixed on the future. Like almost everyone at NIMH, I was enthusiastic about the incredible promise of genetic studies for unlocking the secrets of diseases like schizophrenia. Knowing where genes are located, how they work, and how they send information into cells and tissues would dramatically advance every scientific field, including the study of mental illness. And, indeed, mental-health researchers were beginning to discover thousands of risk-carrying genes in people with various mental illnesses.

In 2013, I am named director of the brain bank, and I quickly settle into this exciting new phase of my career. My work with rat and human brains has long since given me widespread recognition among my colleagues. Indeed, it's what put me on the path that, twenty years after my first paper on the subject, has landed me in charge of so many precious human samples.

Despite multiple discoveries in mental-health research, scientists don't yet completely understand what isn't working in the brains of people with mental illness, and determining how to fix it will likely take many decades and require tenacious dedication from every researcher involved.

And so, despite my brushes with cancer, I work hard, publishing scores of scientific articles and sharing my findings with hundreds of other investigators as we all tackle questions about abnormal genes and the problems they create.

A naturally high-energy person, I bike twenty miles to my office, work all day, then cycle back to our quiet house in the suburbs. Every night at dinner, Mirek and I sit on our elevated back porch as if we are on the deck of a ship sailing through a green sea of woods and grass. We revel in the many birds around us: huge woodpeckers with red caps, tiny house wrens building nests in our flowerpots, colorful hummingbirds feeding on our red impatiens. We feel exceedingly content with life.

Everything seems to be going right—but very soon, I will begin to wonder whether the rats from my early experiments are exacting their revenge on me. Because the same brain structure that I sabotaged in thousands of rodents will begin to malfunction, spectacularly, in my own brain. The cause will not be a neurotoxin injected into my hippocampus that damages my frontal cortex. It will be something far more prosaic, and far more familiar: cancer.

2

The Vanishing Hand

A t the beginning of January 2015, roughly two and a half years after handling my first human brain, I decide to fulfill a dream I've held for years: to compete in an Ironman Triathlon. Though I've completed several Olympic-distance triathlons, I've never tried anything as challenging as an Ironman, which is 140.6 miles of combined swimming, running, and cycling. But it's now or never, my last reasonable chance before I'm too old. I plan to train with a coach and compete this summer or fall in a half Ironman, with three stages that cover a combined distance of 70.3 miles. If that goes well, I will attempt a full Ironman the following year, when I'm at the ripe old age of sixty-five.

It's going to demand incredible effort, but I feel ready and the time feels right. Mirek and our two children, who followed me from Poland some twenty-six years ago, have long since settled into our new home, making wonderful lives for themselves in America just as I have. They, too, are successful and happy. Mirek is a computer engineer in a large software company; Kasia is an endocrinologist at the Yale School of Medicine, where she focuses on diabetes; and Witek is a neuroscientist in the Brain Modulation Lab at the University of Pittsburgh. Both of my children are in happy relationships, and Kasia and her husband, Jake, have two young sons, our beloved grandsons, Lucian and Sebastian, who are growing up rapidly. Mirek and I are celebrating thirty years of a good marriage.

With my family happy and my career going so well, I can devote more time to my hobbies, especially sports. I'm obsessed with developing lean, powerful muscles, not only to feel healthy and strong but because I like looking healthy and strong too. I'm in excellent shape and eager to become even more athletic as I prepare for my greatest physical challenge yet.

In the first days of the new year, I hire a coach and start preparations for the half Ironman. I buy my dream bike, a white carbon-fiber Cannondale Evo road bike with high-end components: eleven-speed Ultegra and deep carbon wheels. Since swimming is my slowest event, I decide to concentrate over the winter on my swimming technique. Several times a week, I get up before dawn and go to a nearby pool to swim eighty to a hundred laps—about two to three thousand yards—before heading to work.

On a Thursday morning toward the end of January, as I pull myself from the pool after one of my first training sessions, I suddenly feel dizzy.

I must have overtrained or run out of calories, I tell myself.

I'm looking forward to a productive and upbeat day. Tomorrow morning, I'm leaving for a conference on brain research in Montana, where I'm meeting Witek and his girlfriend, Cheyenne, for work and skiing, and I'm excited about the trip. But as I drive to work, I have a strange feeling that something's off. My driving feels shaky, although I can't tell what's wrong.

At my office, I sit down and begin to eat a bowl of steel-cut oatmeal I brought from home. I reach out to switch on my computer.

My stomach clenches.

My right hand is gone.

I can't see it. It's disappeared.

I move my hand toward the left.

There it is! It's back!

But when I slide it back to the lower-right quadrant of the computer keyboard, it vanishes again. I repeat my movements and the same thing happens. Whenever I place my hand in the lower-right quadrant of my field of vision, it disappears completely, as if it were cut off at the wrist.

Nearly paralyzed with fear, I try again and again to recapture my disappearing right hand. But once it enters that part of my visual field, it's gone. It's like a freaky magic trick, mesmerizing, frightening, and totally inexplicable—except . . .

Brain tumor.

I immediately try to push the thought from my mind.

No, I think. *That can't be it. This cannot be happening.*

I was sure that I beat stage 3 breast cancer in 2009 and stage 1B melanoma three years ago. But breast cancer and melanoma often metastasize to the brain. I know that a brain tumor in the occipital lobe, the area at the back of the brain that controls vision, is the most likely explanation for this bizarre vision loss. And I know that any brain tumor indicating metastasis—the spread of cancer cells— is horrifying news.

A brain tumor would be too cruel and too deadly, so it *must* be something else. Perhaps the side effect of an antibiotic I'm taking for an infection. I quickly Google *doxycycline,* and sure enough, vision problems and hallucinations are a side effect—very rare, but documented nonetheless.

Clearly, I tell myself, *that's the problem.*

Greatly relieved, I head to the conference room where I'm meeting with a small group of visiting scientists. Once everyone has arrived, we begin to discuss our findings on how genes operate in the prefrontal cortex of patients with schizophrenia.

But I can't focus on the presentation. Whenever I look at the projection screen or at my colleagues' faces, parts are absent, like a surrealist painting or a puzzle with a missing piece. Although the missing section is less than a quarter of my field of vision on just one side, the void still terrifies me.

It feels like there's a hole in my mind. With a terrible gravity, it pulls me back toward the one explanation I don't want to consider:

Brain tumor.

I desperately try to pretend I'm participating in the meeting. But the idea has become stuck in my head: *Brain tumor. Brain tumor. Brain tumor.*

After an hour of torture, I abruptly leave the conference room and run back to my office. I sit at my desk for a while, leaning my forehead against its cool surface as I try to process this bizarre situation. But although I turn it over and over, inspecting it from every possible angle, this symptom has only a single likely explanation—the one that frightens me most.

I have to get out of here. I have to go home. I run to the parking garage, find my car, and drive fast to Annandale, my heart racing the entire way.

At home, my skis and helmet are ready, and my suitcases are packed. I throw a last glance at my notes and piles of conference materials, making sure I have everything I need. Early tomorrow, I fly to Big Sky, Montana, to the annual Winter Conference on Brain Research. As the elected president of the conference this year, I've been key in organizing the meeting, which will draw five hundred neuroscientists from all over the world. I'm also giving the welcoming address to the group, and I have carefully prepared my speech.

I've attended this meeting every year for the past twenty-four years. With its balance of work and outdoor fun, it's my favorite conference. Early each morning, we attend sessions on topics related to brain function, mental illness, and drug addiction. We break for several hours to hit the ski slopes, chatting with our colleagues about our research as we ride the chairlifts up the mountain. Midafternoon, we reconvene for professional sessions and often work together late into the night.

I'm particularly excited this year because my son, Witek, will be at the conference. He and I will work together, then go skiing with Cheyenne. The forecast is excellent—snow for the next five days —and I can't wait to hit the trails. I can almost smell the frosty air and feel the bite of freezing wind on my face as I speed down the slopes, zigzagging through the trees and kicking up blinding clouds of snow.

I love skiing even more than science. It gives me the feeling of weightlessness, an extraordinary lightness of being, a sense of freedom as I fly in and out of control. It's challenging and risky. Navigating the tight trees on a fast run or jumping off the rocks into the white nothingness requires instantaneous decision-making and trust in one's agile body, sharp vision, and strong muscles. And the beauty of the surroundings! Skyscraping mountains over and around me, glittering snow under my feet—the sweet, sweet feeling of paradise.

But this problem with my eyesight weighs heavily on me. I still can't see anything that wanders into the lower-right quadrant of my field of vision.

I try to squelch the panic that is growing inside of me. I just can't accept that this weird phenomenon is serious enough to keep me from Montana. It simply must not be the one thing I've suspected from the moment my hand disappeared this morning, the absolute worst possibility. I won't even let the word *tumor* emerge from my mouth.

But on some barely conscious level, I know my situation may be dangerous. I have to act, and quickly. I call our family doctor, Eugene Shmorhun, and ask for a last-minute appointment. It's late afternoon and near the end of his office hours, but he agrees to see me immediately. I don't tell Mirek or anyone else where I'm going, not wanting to alarm them—and not wanting to admit the terrible possibility even to myself.

Dr. Shmorhun has been our family doctor for almost twenty-six years, since we first moved here from Poland. When we became his patients, he was young, tall, and handsome, just starting his private practice. Over the decades we've all aged together, witnessing one another's skin sag a little and bodies grow rounder. We've joked about our worsening hearing and eyesight. Like us, Dr. Shmorhun likes to run and cycle, and we often discuss with him the results of our latest races. We feel a close connection with him.

Over the years, Dr. Shmorhun has saved our family from a variety of mini-disasters, like my herniated disc and the clot in my husband's subclavian vein that led to the removal of two of his ribs. He was with us when I had my first bout with cancer, a battle that cost me my left breast. Then, in late 2011, he found a melanoma on the skin behind my ear that my dermatologist had missed. My first husband died of melanoma, so I was terrified by the diagnosis, but Dr. Shmorhun saw us through that storm too. Since then, I've allowed myself to become optimistic about my health, and my family has followed my lead. Until today, I was sure that the worst was over. After a painful operation and radiation treatment that beat the melanoma into remission, I was warned by the oncologists that there was a 30 percent chance it would return. But I shrugged off their words. *No way,* I thought. *It's never coming back.*

But as I sit before Dr. Shmorhun and describe my vision problem, my confidence wavers.

"It is the eye, it must be the eye," I tell him. The problem *can't* be with my brain.

As he examines me, I start to speak faster. "I'm taking doxycy-cline, which can cause this side effect," I blurt. "I Googled it."

Hurry up, I think, *I have no time to waste! I'm leaving tomorrow morning for my wonderful trip. Let's get this over with, and fast.*

Dr. Shmorhun continues to check my vision, my eyes, my neuro-logical responses. I notice his serious expression, his unsmiling face. His usual composure is cracking.

"Why worry?" I reassure him. "Things like this can happen."

"I don't think it's your eye," he says.

I freeze. I know that if it's not the eye, it's the brain.

"You can't see anything on your lower-right quadrant with both eyes open, nor can you see that area with either your left or right eye alone," he says. "But your eyes see perfectly well anywhere else. This suggests that your eyes and optic nerves are probably fine but the brain regions that process visual information from your lower right field are experiencing some trouble. I want you to see an ophthalmol-ogist immediately." He leaves the room to call her.

I am terrified.

We need our brains, as well as our eyes, in order to see. The eyes pick up visual information in the world, and the optic nerves send it to the occipital lobe, or visual cortex—the part of the brain where it is processed. If there's a problem in your left eye, you won't be able to see on the left. But if there's a problem in an area of the visual cortex in your brain, neither eye will be able to see a particular visual field—which is the very problem I'm having.

I call Mirek and Kasia and tell them I'm at Dr. Shmorhun's office because I can't see things in the lower-right side of my visual field. Kasia is clearly concerned but I insist it's not a big deal. I say I'll call again after I talk to the ophthalmologist.

The ophthalmologist, Dr. Julie F. Leigh, is right across the street. She checks my vision, dilates my pupils, shines a strong bluish light deep into my eyes. Her pretty young face is close to mine across the slit lamp, her glittering earrings almost touching my ears and cheeks.

I like how she smells, a delicate fragrance of perfume. She finds nothing wrong with my optic nerves or retinas, no cataracts. But when she leans back, her smile has disappeared and her eyes are sad.

"I am afraid it's in your brain," she says. "It must be something in your occipital cortex. We need to do more tests."

I run back across the street. Dr. Shmorhun's office is now closed but he's waiting for me in the darkened reception area along with Mirek, who's just arrived.

Mirek's quiet presence always calms me. Though he was stricken with polio at eighteen months of age and still walks with a significant limp—the polio vaccine wasn't available in Poland until the late 1950s, a few years after it came out in the United States—he is an excellent cyclist with strong muscles in his arms and in his dominant leg. He's an intellectual, unfailingly kind and warm, with a wry but gentle sense of humor. I have a strong personality, loud and laughing and stubborn about my opinions, but Mirek loves me just as I am and is always supportive of whatever I want to do.

I look to him for comfort now, even as I stand defiantly apart from him and Dr. Shmorhun in the dark waiting area. My brave exterior is beginning to crumble.

"We have to do an MRI of your brain as soon as possible," Dr. Shmorhun says.

"But I'm leaving tomorrow morning! I have plane tickets!" I respond. "I'm the conference president, I have to go!" The words pour out in a terrified stream. "I have to go, I have to ski, there will be no conference without me, I am *essential!*" I repeat the same points over and over, like a child desperately trying to convince her parents to let her stay up past her bedtime.

Dr. Shmorhun is usually subdued but today he is very firm. "I can't let you go anywhere before we figure this out," he says. "It could be dangerous to travel. We need to do an MRI immediately. You need to find any place that can take you tomorrow morning." Mirek sides with him.

I continue to argue for an hour—I'm not one to give up easily on what I want. But they won't budge, and I finally give in.

Okay, I tell myself. *I will do the MRI and delay my trip by one day, just to make them happy.*

Mirek and I head home in separate cars. I follow him closely because my vision loss makes it very difficult to drive. It's dark and I have a hard time navigating the winding, wintry roads. Try as I might, I can't stay in the middle of the lane.

When we get home, I call the airline and postpone my flight by a day. I also call Witek to say he should still go to Big Sky but I'll be late joining him. Tomorrow, January 23, is his birthday, and I feel terrible I won't be there. I call a few friends who are headed to the conference. "You wouldn't believe what's happened!" I say in a cheery voice. "I cannot see well and have to have it checked out before I join you. I'll just be delayed by one day." I try not to let my fear enter my voice.

Early the next morning, we go to a nearby imaging center for the MRI. I insist that I do the driving because I always drive and I want us both to feel that everything is normal. But I drive very poorly, weaving across the lanes. "I'm fine!" I snap, nerves frayed, when Mirek asks to take the wheel. "Leave me alone!"

Somehow, we arrive at the MRI center without getting into an accident. The front-desk person checks me in. Only then does it really hit me that I am being scanned for a possible tumor in my brain.

I am nauseated with fear as I prepare for the MRI, which is going to create a very detailed image of my brain and perhaps reveal some horrifying things. A nurse inserts an intravenous line in my arm, which delivers into my bloodstream a contrast liquid that is absorbed by brain tissue. The MRI will use a computerized system to produce pictures (or scans) of my brain that doctors will examine for tumors, strokes, nerve damage, and other abnormalities that x-rays, CT scans, and ultrasound machines cannot reliably detect.

A technician slides me into the tight tube of the MRI machine and turns on the noisy magnet. I lie motionless for an hour before the scan is complete and I am finally free. When we head home to wait for the results, Mirek drives. I am completely exhausted, drained by fear and the stress of the scanning procedure and what it might reveal.

We are home by midmorning; my flight leaves this afternoon. I pack and repack, adding this and that: an extra pair of warm gloves and socks, the sunscreen that I almost forgot. I am hoping the doctor will call soon with the only possible news—that it's not a tumor.

But the impossible happens.

At around 11:00 a.m., the phone rings. I pick it up and sit down on a stool as Mirek runs to join me in the kitchen.

"I am so sorry," Dr. Shmorhun says. "I don't know how to even tell you this." His voice breaks; he pauses. "The scan found three tumors in your brain," he finally continues. "You have to go to the ER right away. One tumor is bleeding, which strongly suggests it might be melanoma. Melanoma tumors have a tendency to bleed. It can be very dangerous."

Watching my face, Mirek knows our world has taken a tragic turn.

I think about the weather.

It is a bright, sunny day here in the Washington suburbs. Snowstorms are in the forecast for later today and tomorrow. And it's going to snow in Montana.

I try to stand up from the kitchen stool but can't move.

I am going to die.

For a fleeting moment, that thought floods through me. But I kick it away with all my might and spring into action. My response to emergencies of any sort is to throw myself into a rational, organized plan and grasp whatever control I can.

I hang up with Dr. Shmorhun and immediately telephone my son. "Witek, I cannot go to Big Sky. I have tumors in my brain," I say. "I am so sorry. It is your birthday, and I am not going to make it." He is, of course, shocked, and I feel like a bad mother for putting my family through so much pain again. I telephone Kasia in New Haven and my sister, Maria, in Boston. Both are stunned. I call my colleagues at the conference and suggest they ask a past president to substitute for me and deliver my speech, which I will e-mail to them. They, too, are dumbstruck.

For my sake and my family's, I am determined to get the best care possible, and I begin to research my options. Staying focused on a plan of attack keeps me from obsessing over the tumors that, at this moment, are flourishing in my brain.

I call Dr. Claudine Isaacs, my breast cancer oncologist at Georgetown University Hospital. "A horrible thing has happened. I have tumors in my brain," I say. "Maybe it is breast cancer metastases. But one tumor is bleeding, so my family doctor thinks it is melanoma. Where should I go?"

When she speaks, it is clear she is shaken. She tells me to go immediately to the ER at Georgetown and directs me to see Dr. Michael B. Atkins, a melanoma oncologist who she says is phenomenal. She says she will meet me there.

From the corner of the hallway, poised for my trip, my skis stare at me—sleek, beautiful Rossignols that I bought last year. They respond to the slightest movement of my feet, my toes—even my mind, it seems. With them I fly through the snow, fluid and graceful. Now I'm headed to the hospital, and they will have to stay behind.

It's Friday afternoon before a snowstorm, not a good time to enter an emergency room. My blood pressure is sky-high, perhaps from anxiety, perhaps from a bleeding brain tumor. The nurses give me steroids to prevent brain swelling caused by tissue irritation from the bleeding tumor. I lie for hours on a cot behind a flimsy curtain. All around Mirek and me echo the sounds of rushing, crying, screaming; the sounds of distress and endangered human lives. It's devastating to be back in this world just three years after undergoing surgery for skin cancer.

Doctors come and go, all asking the same questions, and I tell them the same thing: "I cannot see on the right lower side. My MRI shows brain tumors, and one is bleeding. I've had breast cancer and melanoma."

It turns out that Dr. Atkins is away today but Dr. Isaacs comes in and offers words of support. She leaves. More doctors cycle through the room. A neurosurgeon swings through and advises against brain surgery in favor of radiation, which will be safer than cutting into my brain. A radiation oncologist visits and gives the same recommendation. No decisions are made. We wait for hours.

Maria calls again and again from Boston, where she is a physicist and chief of therapy in the radiation oncology department at Brigham and Women's Hospital.

"Come to the Brigham," she insists. "The doctors here are the best. I talked to Dr. Aizer, a radiation oncologist. He says that surgery should be done first and then radiation."

How can I possibly go? I'm lying here in the ER with a bleeding tumor in my head. Despite all my years of studying the brain, I'm not a neurologist or any other kind of medical doctor. I know close to nothing about what could happen to me. Will the tumor burst open and flood my brain with blood? Wouldn't that kill me? I'd better not move. But Maria wants me to see the doctors she knows and trusts. What should I do?

Shortly after 8:00 p.m., the flimsy curtains part and Witek and Cheyenne appear. They canceled their trip to Montana and drove down from Pittsburgh. Oh, what a joy it is to see them! Despite my fear and despair, I'm ecstatic that they're here. Soon after, Kasia arrives. She took the Acela train from New Haven and made it just before the storm. Mirek and I are so happy to have everyone together, to breathe the smell of their bodies and touch their faces, kiss their cheeks. Kasia is very tired; a few hours earlier, she herself was seeing patients. She lies down with me on the cot and we snuggle closely like we did when she was my little baby. Witek and Cheyenne fetch sushi from the hospital cafeteria, and we share a feast on my bed amid the

IV lines and crumpled sheets. We're surrounded by the frightening sounds of the ER, but we're together in this ordeal, my family and I.

At midnight, they leave. I remain in the ER listening to beeping and more beeping, to the tragic noises of people in desperate need of help. Nurses peek in from time to time, and I plead with them to transfer me to a quieter place. At three in the morning, they move me to a room in the ER that I share with an older woman who is in serious pain and surrounded by a large family.

In the morning, Mirek and my children return, and our waiting resumes. It's Saturday, and the hospital is overcrowded. No doctors stop in to see me. Nothing is happening. By noon, we've made our decision—we are leaving here and going to the Brigham in Boston tomorrow. But it isn't as easy as we thought. The attending physician refuses to approve it, and the nurse tells us that insurance will not pay for the ER visit if I check out against medical advice.

"I'm afraid to leave without their approval," I tell Kasia. "What if the tumor bleeds even more? And this hospital visit will cost us tons of money if insurance doesn't pay!"

But Kasia is checking on her iPhone for the patient's bill of rights and the insurance rules, which contradict the nurse. "It's not true," Kasia says. "We are leaving, Mom."

We head north to Boston early the next day, Sunday, January 25. Before we leave, my friend Jania, a hairdresser, comes to my home to cut my hair. I called her at dawn and told her my news, and she rushed over, arriving in her pajamas at 7:00 a.m. I ask her to give me a crewcut in case they open my skull.

"It will be easier for the wound to heal," I explain.

Mirek and I pack our Toyota RAV4 with our trainers and road bikes so we can use them as stationary bikes in my sister's basement. We agree that no matter what happens, our athletic training can't stop. I also take my skis. Just in case.

Mirek, Kasia, and I hit the wintry roads, snow falling lightly, as Witek and Cheyenne follow in their car. We pass a nearby construction site, where a Giant supermarket is being built. I have been so excited in recent months that we will finally have a decent grocery store in the neighborhood and we'll no longer have to drive for miles in traffic just to do our shopping.

Will I live to see it open? I wonder.

I feel the urge to talk, to plan the future for my family. I'm sure I'm going to die—not right away, but soon, maybe in a few days or weeks. I have, of course, researched my condition on the Internet. The prognosis for metastatic melanoma in the brain is terrible, especially if you are over sixty and you have three or more tumors. I have three tumors, and I'm sixty-three years old. Four to seven months of life is all I've got left. I will be dead as early as May, as late as August. I won't make it to sixty-four.

As I sit next to Mirek, who is driving, I can't stop thinking about my family's future. I need to write my will and create a trust for my assets to make things easier for them. I want my belongings to be fairly divided with no arguments, no lawyers, no complications.

"Mirek will have to sell the house," I say to Kasia, who is sitting in the back seat. "He needs to move closer to you kids or my sister."

"Stop, Mom," Kasia says. "Let's talk about something nice. We'll go cross-country skiing. You'll like it." I stop speaking about my plans because I see my grim preparation is hurting them. But I continue silently.

Mirek cannot stay alone. How difficult would it be for him in our house, with everything the same but without me there anymore? How would I feel if he were gone? How lonely to come back to a darkened house—my clothes still there, my earrings, my life as I had left it. But no me.

I feel so sorry for him that my eyes well up with tears. I'm afraid they will see me crying. I have to shake it off and stop thinking such thoughts. But Kasia knows. "Mom, it'll be all right," she says tenderly.

"Mirek will be fine. We'll all be fine. Don't worry." But of course I worry. I worry for them, and for myself.

We all stop for the night at Kasia and Jake's house in New Haven. Our grandsons, Lucian and Sebastian, greet Mirek and me with shrieks of joy. They don't fully understand what's happening but they know that Babcia (Polish for "grandmother") is sick and everyone is worried.

This house is freighted with meaning and memories. When Mirek, Kasia, Witek, and I first moved to America in 1989, we lived in a rental apartment in a subdivision of townhouses in Alexandria, Virginia, among a sea of immigrants from around the world. We were delighted by the size of our apartment. With bedrooms for each child, it was the largest place we'd ever lived and seemed like a mansion. We owned no furniture, so a work colleague loaned me a queen-size air mattress that Mirek and I shared, and the children slept on large pieces of foam we bought at a garage sale for a dollar each. At a church sale, we paid thirty-five dollars for a chrome-plated table and beaten-up chairs with plastic yellow cushions that felt luxurious after weeks of sitting on the floor and using a cardboard box for our table.

It was Kasia who first mentioned that the only kids who got off the school bus at the housing complex were recent immigrants. Other kids—she meant *richer* kids—lived in single-family homes in nice neighborhoods. We researched what it took to buy a house and found that a mortgage would cost about the same as our rent. And the money would go into our own property! It was a revelation. The concept of owning a house was thrilling and totally foreign. We began looking for something that we could afford, and in the real estate section of the *Washington Post,* we found a house in Annandale, Virginia, in a neighborhood very close to the one we were in, with large Colonial single-family houses and meticulously kept yards. The property we bought stood out as long neglected, with spots of bare earth and huge tree roots cutting through the front yard, and the house needed a lot of work. But it backed onto woods and a stream. Most important, it was our land, all the way to the center of the earth. We loved the sense of freedom and independence that it offered. It told us that we had made it in America.

Both Kasia and Witek now have beautiful three-story homes of their own. Witek and Cheyenne live in a bohemian district of Pittsburgh, and Kasia and Jake's home is a sky-blue Victorian on a quiet street a mile from the Yale campus. Every time we visit them, the sight of all they have accomplished makes my heart swell with pride and love, as do Kasia and Jake's adorable children, my grandsons, Lucian and Sebastian.

Everything about these boys makes me deliriously happy. The smell of their hair and skin is intoxicating, overpowering. I love their smiling faces, their funny, uneven, oversize teeth, their messed-up, sweaty hair, the energy bubbling in their little bodies. There's nothing I love more than visiting them and playing their games with them, reading to them, walking them to school. I try to cherish every moment of their childhoods—a time in their lives that will, of course, pass too quickly.

Where does it come from, this overpowering love of a grandmother for her grandchildren? Forty years ago, when Kasia was born, my mother-in-law laughed and wept and doted on her first grandchild, clapping with joy and excitement at every minute change of expression on the baby's face, every movement of her tiny hands or feet. I was embarrassed for her. Then Kasia's son Sebastian was born, in 2006, and I quite similarly became a doting grandma. When Lucian was born three years later, it happened again; I felt the extraordinary emotions triggered by becoming a grandmother. Just as my own *babcia* adored me and showered me with unconditional love, I discovered that a grandma's love is boundless, stupefying, sentimental, and capable of turning one's brain into schmaltzy mush. It's also superbly gratifying and blissful. And I have never felt more desperate for these two small, precious boys than I do now.

We all take Sebastian and Lucian to school the next morning, a Monday. It hits me that I may never see them again, and a wave of maddening sorrow rises inside my chest, floods my body, and chokes my throat. I kiss their heads, smell their hair, hug their thin, little bodies, and leave.

Mirek, Kasia, Cheyenne, Witek, and I continue north, leaving Jake behind to care for the boys. He will join us later. It's snowing again as we pass through the stark, binary landscape: white roads, white fields sliced with black rivers, black tree trunks with branches like pencil strokes on white paper. A frozen world.

I feel I am frozen too, as fragile as a thin sheet of ice. A tap in the wrong spot and I could shatter.

We arrive in Boston before noon. Maria has already arranged for appointments today with various doctors at the Brigham and Women's Hospital and the affiliated Dana-Farber Cancer Institute. My melanoma oncologist, Dr. Stephen Hodi, is at Dana-Farber, while Dr. Ayal A. Aizer, a warm, caring, and meticulous radiation oncologist, and Dr. Ian Dunn, a neurosurgeon, are on staff at the Brigham. They will all work together on my case.

There are six of us at each appointment—Kasia, Witek, Cheyenne, Mirek, Maria, and I—as well as the doctor and a nurse and sometimes a resident or an assistant too. Sometimes the doctor has to ask which of us is the patient, which amuses us. We crowd the rooms, my tall and handsome family—my sister and I are the smallest—and at each visit, the staff has to add chairs.

Each doctor does the same simple test of my vision: he raises the index and middle fingers of one hand in a V shape and moves the V up, down, left, and right, in each of four visual quadrants, asking whether I can see it. When the V is moved into my lower-right quadrant, it's invisible to me.

I immediately have another MRI as well as a CT/positron emission tomography (PET) scan, which will reveal the location of any fast-dividing cancer cells. We spend a long time with Dr. Aizer, who explains why surgery on the bleeding tumor should be done first, followed by radiation of the area and of the two other tumors. He takes care to describe everything clearly and spends hours discussing the scans. My oncologist Dr. Hodi, a world-famous expert in cutting-edge treatments for melanoma, says surgery and radiation must be done before he steps in with other therapies. His explanations are convincing, and we all agree with his proposed treatment plan.

As we wait to meet the neurosurgeon, Kasia looks at my records and exclaims, "Oh my gosh! Your surgeon is Ian Dunn, my friend from medical school."

"Is he any good?" I ask.

"Fantastic!" she assures me. "Very studious."

My family crowds into the small office, and Kasia is sitting with me on the examining table when Dr. Dunn arrives with his assistant. He and Kasia chat and laugh. "What a coincidence!" he says.

Dr. Dunn pulls up the scans on his computer and points to the frightening shapes on them. I glance at them quickly, then look away; as much as I have studied brains, I don't like peering into my own when it is damaged like this. I don't like to see scary black spots where healthy gray tissue should be.

Just as my ophthalmologist and I suspected, the tumor causing my symptoms is in the primary visual cortex, in the occipital lobe at the back of my head, which is why it's affecting my vision. The size of a large raisin, it's nestled down in the sulcus, the narrow valley between two gyri, like a little black sheep hidden in a crevice between two hills. Although it's bleeding, it's not in the worst place, I tell myself. If it were in my spinal cord, I might be paralyzed. If it were in my brain stem, which controls basic life functions like breathing, then surgery might be too dangerous and out of the question. I'm fortunate that it grew in a place that didn't threaten my life but let itself be known. Had the tumor developed without noticeable symptoms —if my hand hadn't disappeared and freaked me out—it might have thrived for quite a while before any of us noticed anything wrong. I'm sure I would have died. There's a lot of luck in this unlucky situation. This nasty little raisin is saving my life. For now.

Dr. Dunn explains that he will stop the bleeding and remove the tumor. A lab will examine it to determine whether it is, in fact, melanoma, and if so, what kind.

"Am I going to go blind?" I ask. Surgery always involves serious risks, including, in my case, damage to the occipital lobe, which could result in the loss of vision.

"Probably not, although theoretically, it's possible," he says. "And if you don't, you may still have vision problems. It's also possible that you won't wake up after surgery. That's unlikely, but I must inform you of all the risks."

His young male nurse, energetic and cheerful, presents a consent form listing all the terrifying things that could go wrong. I sign it and we leave.

The surgery is planned for the next day, Tuesday, January 27. But a huge snowstorm is on its way. It will come to be known as the Blizzard of 2015, a nor'easter that will dump tons of snow in the northeastern part of the United States and Canada. As we drive to my sister's house in the Boston suburbs, snow is already falling. The narrow, winding roads are slippery and soon become shrouded in white. The Toyota skids often as we hold our breath.

We end up waiting two more days for my surgery as the blizzard covers the world around us. The snow piles up to the windows of my sister's house. After the storm, it's beautiful outside, quiet and calm. I walk with Kasia and Witek in the woods, and the snow reaches up to our thighs. It is light, fluffy. I lie down on my back and make snow angels. We are laughing. It is so good to be alive.

Since the surgery is delayed, I spend my time enjoying my family, and I completely block out any thoughts about the tumors. Although I'm an expert in the brain, I'm repelled by what's going on inside my own.

When I held that first brain in my palms at the brain bank, I could admire it with detached interest—because it wasn't mine. Now, while I want to participate in my treatment by selecting a highly skilled team of doctors, I don't want to look at my MRIs or think about what's happening in my skull. My own brain presents a mortal danger to me.

It is Thursday before the roads are clear enough for us to make it back into Boston.

The traffic is heavy that morning, and it takes forever to get to the hospital. The streets are clogged with cars moving very slowly in deep snow, and there's more snow in the forecast. Finally, we arrive. My whole family is with me, including Jake, who's joined us after leaving the boys with his mother in New Haven.

In the late morning, we enter a large area with semiprivate cubicles furnished with couches and comfortable armchairs; these provide families some privacy as they wait for their loved ones to come out of surgery. My family has brought all kinds of things to entertain themselves with: books, games, computers. They were told the wait could be long—the blizzard may have contributed to the delay—and two or three hours pass before I'm even taken back to the pre-op area. But we are all in good moods, joking and chattering as if we are at a party, abuzz with nervous energy.

When they call for me, I head into pre-op with Mirek and my sister. There, I'm examined by the nurse, meet the anesthesiologist, and visit again with my surgeon Dr. Dunn. Far from being scared, I feel immense relief that the surgery is finally happening, that I soon will be under anesthesia and won't know or remember anything.

A nurse administers a strong sedative as I sit in the pre-op room and soon I begin to float away. I embrace the darkening of my mind, unaware that this brush with oblivion is only the beginning of my long and dangerous journey.

3

Into My Brain

As soon as I am unconscious, Dr. Dunn drills into the back of my skull to reach the bleeding tumor in the occipital lobe. He finds the nasty raisin relatively easily; it's growing between the folds of my primary visual cortex.

With the help of his surgical team, Dr. Dunn scoops out the tumor and suctions out the blood. He replaces the portion of my skull that he removed to gain access to my brain, seals the bone with titanium screws, and stitches me up. To keep the sutures intact, he folds the skin of my scalp and rolls it along the five-inch incision so that it looks like a fat earthworm glued to the back of my head. Later it will flatten out into a neat scar.

A few hours later, I open my eyes.

The first thing I notice: I can see! I'm not blind! I can see everywhere, in all my visual fields—left, right, up, and down. I gaze around the hospital room testing my eyesight, raising my fingers in a V shape and moving them into each of the four visual quadrants like my doctors did before the surgery. No problems; none! I can see the V no matter where I place it! No vanishing hand, no blocked fields, nothing abnormal. The tumor and bleeding haven't caused permanent damage to my occipital cortex.

I'm so relieved—but for one detail.

Dr. Dunn informs us that the tumor appears to be metastatic melanoma. We'll know for sure in a few days after we get the results from the lab. In the meantime, all we can do is stew over the idea that, just as we feared, I am most likely battling this dreaded type of cancer all over again.

Melanoma is the rarest but most dangerous form of skin cancer, diagnosed in about 130,000 people each year, most of them fair-skinned like me. It develops from melanocytes, skin cells that carry a dark skin pigment called melanin, which protects deeper skin layers from the damaging effects of the sun. Many melanomas begin as moles, harmless growths of melanocytes that can turn cancerous over time. Once that happens, melanoma has a propensity to metastasize, often spreading from its original site in the skin to lymph nodes and organs, especially the lungs, liver—and brain. When it spreads there, it's almost invariably terminal.

For all we know, I've been handed a death sentence.

We have no doubt I'm going to die. My family and doctors and I are certain of it. We don't discuss it aloud, but the terrible reality of it lingers among us.

That night, Thursday, January 29, my exhausted family heads to my sister's house while I stay in the hospital to recuperate. As I lie in bed, I feel no pain—but I cannot sleep. I'm loaded with steroids to prevent swelling in my brain, and one of the side effects is insomnia. I'm wide awake, my mind exploding with memories.

In this dark hour, the intensive care unit nurse who is monitoring me pulls up a chair next to my bed and sits down. As the snow falls outside the window, words spill out of me. I tell her things I've never shared before, painful stories I thought I'd left behind in Poland. I talk all night.

The next morning, Witek and Cheyenne are the first to arrive. In the quiet of my hospital room, I share these stories with them too. I'm sure I am going to die, and I want them to know my history, which is their history as well. I especially want Witek to know more about his father, Witold, a brilliant computer scientist.

I also have a selfish reason for telling these stories: I need to express my fear about what is happening within my body, to give voice to the family history that is now repeating itself in the most painful way. Because when my son was only seven years old, his father died of the very type of cancer I now appear to have: melanoma that metastasized to the brain.

Witek was still a toddler and his sister, Kasia, a five-year-old when my husband told me the news. It was June 1980, a hot and sunny day in Warsaw, Poland. I was twenty-nine years old, a young wife and mother, cutting up vegetables to make dinner, when Witold walked into our home, his face contorted with fear.

The words that poured from him were so dreadful I could barely process them. Earlier in the day he had gone to the local hospital to see a dermatologist after spotting a dark mole on his back. The doctor had taken one glance and announced that Witold had melanoma.

"He said I'm going to die," Witold said. "Eight months, at best."

I wanted to scream but no sounds came out. Finally, I shouted, "He must be wrong!"

Surely the doctor was just another quack, one of the many awful medical professionals in the health-care system in Communist Poland. One look at Witold told you that he was perfectly healthy. He was handsome, broad-shouldered, and muscular, a swimmer and runner at a time when almost no one else in Poland ran for exercise. We were a beautiful young family with two picture-perfect children.

By Polish standards, we were well-to-do, accomplished, and worldly. We had just spent the 1978–1979 academic year at the University of Illinois at Urbana-Champaign, where Witold had studied on a Fulbright scholarship. We had ambitious plans for our future. Cancer was not among them.

Early the next morning, we rushed to the same Warsaw hospital demanding to be seen. The doctor was solemn and cold as he repeated his original diagnosis: Witold would die within months. "There is no cure," he said. "Prepare yourselves." I felt faint. A nurse pressed a Valium into my palm and ushered us out the door.

"We're not going to tell anyone about this," Witold whispered as we lay in bed that night. In Poland at the time, cancer carried a stigma. Even among our enlightened and educated friends, it was viewed as a sign of weakness and loss of control over one's life. Talking about it was taboo.

A few days later, an oncologist confirmed that Witold had melanoma and scheduled immediate surgery. Within weeks, the melanoma had been excised and my husband had begun chemotherapy.

The infusion unit in the Institute of Oncology on Wawelska Street in Warsaw was frightening and depressing. To make matters worse, we—like most people at the time—knew almost nothing about chemotherapy. No one told us what to expect or what the treatment was intended to accomplish. Doctors and their staffs didn't communicate with patients, and families were left completely to their own devices. In those years before the Internet, there was no easy way to get information. I was very aware, however, that our situation was grim. Cancer, especially melanoma, was considered a terminal disease. Very few lived through it.

But the weeks passed and Witold did not die. After surgery and several rounds of chemo, he returned to his normal life, and quickly, I started to forget that his cancer had ever appeared. I did more than forget; I deliberately kicked his illness out of my consciousness. I shoved it into a dark corner of my mind, covered it with layers of superficial happiness, and nailed it shut with vodka and partying.

Still, the nightmare of his disease—no matter how deeply submerged in my unconscious—hung over us. Witold grew more and more withdrawn, and in our denial of the seriousness of his illness, we pushed each other away. I was scared, as much as I tried to believe I wasn't. Fear fueled our isolation, and we drifted farther apart.

By the end of 1981, the political situation in Poland had come to mirror my deteriorating marriage. That December, the Communist government declared martial law in an attempt to crush mounting political opposition inside the country, drastically limiting the freedom of Poles and sending the nation's already wobbly economy into a tailspin. The streets of Warsaw were blocked with tanks and patrolled by Polish soldiers in full military gear. On freezing nights, they warmed themselves at makeshift fires that bloomed across the dark city. It was an alien world to us, frightening, practically a war zone. Long lines of people waited for food in front of empty stores, soldiers at checkpoints examined IDs, people rushed home before curfew for fear of being arrested, friends were thrown into jail.

By the time I fell in love with another man, Mirek, my marriage with Witold was all but over. I consoled myself with this fact each time I fell into Mirek's arms; his steady presence was exactly what my children and I needed. Witold took the news of my infidelity hard. He disappeared from our lives by moving to France and he visited the children only a few times over the next two years; crossing back and forth to the West wasn't easy.

On one visit, as he was leaving my apartment, Witold turned in the doorway and told me I was a great mother, that I had always been such a strong force standing unconditionally by our children, and that he envied my conviction and dedication to them. He was sad, warm, and humble. He kissed me goodbye, his first friendly gesture in years.

I had no way of knowing it then, but those would be Witold's last words to me. In May 1985, a few months after that visit to Warsaw, he died in a hospital in Bordeaux, France. The cancer had metastasized to his brain. At that time, there was no cure for that kind of brain cancer.

When I got the news, I shook uncontrollably, and the children wept when I told them. They were too young for a funeral, I decided along with my family and Witold's, so I went alone. Later, when I tried to bring up their father's death, they didn't want to talk about it. Over the years, we simply did our best to move on, each of us in his or her own way. But Witold's death still hovers over all of us, and melanoma carries an especially potent meaning for our family.

By Sunday, February 1, 2015, three days after my surgery, I've healed enough to be released from the hospital. Mirek and I head to my sister's house, where I continue recuperating while remaining close to my doctors.

Still full of steroids to keep my brain's swelling at bay, I feel like a superhero with limitless powers, a wild woman on stimulants—driven, driven, driven. From Boston, I send a series of e-mails to the administrative, clinical, and scientific directors of the NIMH, telling them all the things I want them to know in case I die. The e-mails make sense but there are a lot of them, and they are long and very detailed, a sign of my steroid-fueled manic energy.

I can't stop the flood of thoughts. I can't stop talking and writing. I write pages and pages about my life. I need to make sure that all that I am and have been doesn't disappear should this disease take me. And the chances are very high it will. Despite my physical fitness, my passion for life, and my deep love for the people around me, I am going to die, and probably soon. I know it and my family knows it. Training for the Ironman is over. My life as I've known it is over.

But I am not going down without a fight, and, strangely, I'm feeling optimistic. Since my first husband's death from melanoma, I've kept up to date on the latest research about this terrible disease. Every time I read about another medical advance, I think about Witold and wonder, *What if he'd survived long enough to receive this treatment? Would he still be alive today?* It is heartbreaking to think that the astonishing advances in this field came too late for him.

The newest and most promising front in the battle against cancer is immunotherapy. This cutting-edge treatment uses the body's own defenses to fight the disease, empowering the immune system to recognize and destroy cancer cells that would otherwise evade it. Research organizations, scientific journals, and even newspapers and TV news programs are touting immunotherapy as the most exciting and encouraging advance in cancer treatment in decades, perhaps in all of medical history.

My melanoma oncologist, Dr. Hodi, who treated me in 2012 when they found the melanoma on my neck, is a renowned expert in cancer immunotherapy. Although we're still waiting for the lab results, based on Dr. Dunn's evaluation, Dr. Hodi has no doubt that I have metastatic melanoma. After I recover from my surgery and undergo radiation therapy, we will discuss additional treatment options. Will immunotherapy be among them? This is my fondest hope—but I also know it's a long shot. At this point, in 2015, there are scant reports on the effectiveness of immunotherapy in treating brain tumors, and the latest drugs haven't been applied yet to metastatic melanoma in the brain. For all I know, people like me are doomed.

I could easily despair. But years ago, I learned an important lesson from an unlikely source: Lance Armstrong. In 2007, my father was dying from colorectal cancer, and I flew back and forth from the United States to Poland to take care of him. I did a lot of reading on the very long flights, including, one night, Armstrong's memoir about surviving cancer, *It's Not About the Bike: My Journey Back to Life.*

At the time, my own battles with cancer were still in the future, but I wept as I read Armstrong's book all the same. I identified with his competitive spirit and was deeply impressed with his approach to dealing with his disease, especially when it seemed there was no hope and he was destined to die young. When some doctors gave up on him and he didn't have health insurance or the money to pay for treatment, Armstrong taught himself about his particular type of cancer, testicular cancer that metastasized to his lungs and brain. He then found the best institution and specialists in the United States for treating it.

You must be your own best advocate, Armstrong insisted. You can't rely solely on your doctors or your family or anyone else; you have to stay on top of your own care, no matter how sick or exhausted you feel. Learn everything you can about your disease and your diagnosis, locate the very best doctors, find out exactly what drugs and treatments your doctors are giving you and what they're supposed to do, never stop researching and asking questions, and check, check, check what the doctors tell you—get second and third opinions. All of this is up to you because ultimately no one else—not your family members, who love you, or your doctors, who want you to survive—is responsible for your health. You need a support team, of course, but in the end, you run this race on your own.

The comparison to a race is no idle metaphor. In high-level athletic competitions, as Armstrong wrote, suffering is part of the process. A high tolerance for pain, both mental and physical, gets you across the finish line. As a marathon runner and triathlete, I understood exactly what he meant when I read his book eight years earlier. Now, faced with the most daunting challenge of my life, I know that the athletic competitions I love so much are the very best preparation for enduring what lies ahead and, perhaps, surviving it.

I am getting ready for the competition of a lifetime. I have a high tolerance for physical punishment, and I have trained myself to never give up, no matter what. As I face this disease yet again, now in its deadliest form, that attitude—*I will do this, I will make it*—becomes my lifeline. Excellent health care and unwavering perseverance saved

Armstrong's life. I'm hoping it will save mine. The stakes are the highest they can be. Living—that's the ultimate win.

And so, although the odds of my survival seem slim, my family and I set out to learn everything we can about metastatic melanoma. Luckily, we are well equipped for this project: Witek is a neuroscientist, Kasia is a physician, my sister, Maria, is a physicist working in radiation oncology, and Mirek is a brilliant, logical, cool-headed mathematician. Together, we study the mechanisms of metastatic melanoma and the best treatments available. We scour medical journals for the latest studies, and we visit doctor after doctor.

I am, of course, terrified at the thought of my own death. But I do not allow myself to get depressed. I don't curl up in a ball or cry. That would drain precious energy that I need if I'm going to have any chance of survival.

It's not the first time I've refused to give up easily. Before I began chemotherapy for breast cancer six years ago, an acquaintance called to inform me that the mastectomy would be horribly painful and the chemo would leave me so depleted that I wouldn't be able to move. She said that she was sending me a gift that I would need. A few days later, soft, polka-dot pajamas arrived in the mail along with a note offering her best wishes and telling me to prepare to spend a lot of time in bed.

While I appreciated the gift and the wishes, she couldn't have been more wrong.

After my breast was removed, I did stay in bed—for two or three days. By the fourth day, I was up and walking outside, eager to get back to the business of life. I'd resolved to ignore my pain and discomfort as best I could and focus on recovery. I couldn't stand to look at the pajamas, so I gave them away.

This episode has become a running joke in our family. When I received my new diagnosis, Mirek and my children asked, "Should we send you polka-dot pajamas?" *Not a chance,* I thought.

I have no intention of feeling sorry for myself. Self-pity destroys my composure and sucks up my energy more than anything else.

But I also don't realize how bad things are about to get.

In mid-March, roughly a month and a half after my neurosurgery, a series of MRIs show new small lesions—areas of abnormal tissue—in several brain regions. They are most likely tumors, although it's hard to tell from the MRI alone.

Witek, I, Kasia, and Jake go cross-country skiing outside Boston exactly one month after brain surgery to remove the tumor in my occipital cortex.

Dr. Aizer, my radiation oncologist at the Brigham, believes that stereotactic radiosurgery (SRS) is the best option for the tumors. SRS focuses high doses of radiation onto individual tumors with the goal of withering them into oblivion. Another approach is whole-brain radiotherapy, in which the entire brain receives a somewhat lower dose of radiation. But Dr. Aizer says whole-brain radiotherapy is not the best option for melanoma because high doses of radiation are required to kill these particularly aggressive cancer cells. Anyway, I do not want to consider that sort of scorched-earth approach. Radiation is not a benign procedure, after all; its purpose is to kill cells, and it doesn't discriminate between cancer and healthy cells. The thought of having my whole brain bathed in neuron-destroying radiation horrifies me.

For some patients with advanced brain melanoma and many brain tumors, SRS is not viable—there are just too many sites that would need high-energy radiation. This could lead to dangerous brain tissue damage, which of course worries me very much. Fortunately, at this point I have few enough tumors that the targeted approach of SRS just might work. And so I am strapped onto a gurney with a custom-made facemask to hold my head in place, and we shoot the few small tumors with precise, high-energy radiation beams in the hope that they will wilt into nothingness.

But targeted radiation therapy is not a permanent solution. If new tumors continue to appear—as they clearly are—my brain will soon be riddled with deadly lesions. The doctors will stop the radiation treatment because it will be futile; there is a limit to how much radiation a brain can withstand without permanent damage. The tumors will continue to grow, pressing on my brain and causing swelling in the tight compartment of my skull. Eventually I will fall into a coma, and finally—when the swelling squeezes the brain stem at the bottom of my skull, cutting off my ability to breathe—I will die.

I have to do something dramatic, find something cutting edge that might save my life. Without some sort of novel, more aggressive treatment, I will be dead in a matter of months. My family and I continue to read every new study that's published in the medical journals. We visit melanoma specialists in Boston, clinicians and researchers, gathering information and analyzing their advice. Secretly, I also hold out hope that my melanoma oncologist, Dr. Hodi at Dana-Farber, will recommend some sort of phenomenal new immunotherapy treatment.

Yet when we next visit Dr. Hodi, whom I have not seen since shortly after my brain surgery, he is solemn upon hearing that I have additional brain tumors. To my disappointment, he says he's not sure that immunotherapy is right for me at this time. Doctors don't know yet if it's successful in treating advanced melanoma in the brain, he explains. I suspected as much from my own research. At the end of the visit, Dr. Hodi mentions the opportunity to join a clinical trial in Boston. But I have doubts about whether to go this route, not least because of the difficulty of participating in a trial so far from home.

We really don't know what to do next. So we continue our search and visit Dr. Keith Flaherty at Massachusetts General Hospital, a warm and knowledgeable doctor in a bow tie who spends an hour and a half explaining the new treatments for melanoma. Not only is he a specialist in targeted therapy—a promising treatment that targets specific molecules in cancer cells—but he's an expert in treating specific mutations in melanoma. Despite Dr. Flaherty's experience with targeted therapy, he suggests that I try immunotherapy first. He tells us about a clinical trial in immunotherapy for patients with melanoma brain tumors that is just about to open at Georgetown's Lombardi Comprehensive Cancer Center under the direction of a highly regarded oncologist, Dr. Michael Atkins—the same oncologist my breast cancer doctor recommended when my brain tumors were discovered in January.

"Dr. Atkins is very good. I've worked with him," Dr. Flaherty tells us. "You should be treated there. It will be very convenient since you live in that area, and he's a great doctor."

Given my poor prognosis, my family and I agree that the best approach is to attack the melanoma with every possible weapon: radiation, immunotherapy, and then, perhaps, targeted therapy. "If you get all of those, it'll be like we're throwing the kitchen sink at you," Dr. Flaherty says with an encouraging smile.

In late March, about two months after neurosurgery and after several sessions of radiotherapy, I finally leave Boston and return home to Virginia. The incision on the back of my head has turned into a long scar, which is clearly visible, since my hair, shaved for surgery, hasn't grown back yet.

My new white bike is waiting. Orphaned in a dark corner of the garage, it looks at me reproachfully, as if asking, *Why did you bring me here if you're going to die?* I pat its soft, white handlebar, and, for the first time since this ordeal began, I cry. "I promise I will ride you," I whisper.

A day later, I keep my word. I climb on the bike and start riding slowly through the quiet streets of my neighborhood, steering cautiously so I don't fall and hurt my recently stitched and radiated head.

Roughly two months after brain surgery, I tentatively train in the streets of suburban Annandale, Virginia.

My doctors have told me I have to wait a few weeks after radiation before I can start another treatment. And so, at the end of March, Mirek and I—along with Kasia, my sister Maria, and her husband, Ryszard—escape to the Big Island of Hawaii to forget our dark thoughts of death and to gather strength from one another. Mirek, Kasia, and I bicycle more than two hundred miles through the lava mountains. My vision is perfect, my brain working as it always has, and the fact that I'm not experiencing any symptoms gives me some hope that I'm getting better. I'm filled with optimism. I begin running a few miles every day and working out nearly as vigorously as ever. Out in the open ocean, I swim part of the course of the famous Lavaman Waikoloa Triathlon that is soon to take place. On a whim, I even enter a 5K race through lava fields, and I place fourth in my age group.

Hawaii affords us a blissful respite from the chaos of the past two months. But in the back of my mind I am endlessly turning over Dr. Flaherty's advice. I'm imagining what the immunotherapy trial at Georgetown might be like and whether it could actually work if I'm able to enroll in it upon our return home. If not, what then? Will I be able to run, bike, and swim for much longer? Will I ever see this beautiful place again? And what of my family? Will this be the last happy time they remember spending with me?

Every night in Hawaii, all five of us stretch out on the tropical lawn in front of our bungalow, hold hands, and stare for hours into the enormous, glittering sky. I don't want to die. I lift my foot to touch a star with my big toe, and then another star, and another, and make wish after wish. Soon, five pairs of feet are dancing across the stars, skipping over the vastness from which we came and to which we'll return. We are together now, tight as can be.

When we come back from Hawaii at the beginning of April, I call Dr. Atkins at Georgetown Medical School, which is about twenty miles from where we live. Two days later, Mirek and I meet with him.

Dr. Atkins describes the protocol of the upcoming clinical trial, known as CA209-218, which is taking place in sixty-six locations and will include several hundred people in the United States and Canada. Every three weeks, he explains, a combination of two monoclonal antibody drugs, called checkpoint inhibitors, will be simultaneously infused into my veins to boost my immune system. These drugs are supposed to teach dysfunctional T cells, which are fooled by cancer into ignoring the disease, how to recognize, attack, and (we hope) kill off the melanoma cells invading the body. The drugs, ipilimumab and nivolumab, are used for the treatment of advanced melanoma. They were separately approved by the FDA, in 2011 and 2014, and in a very short time, they've revolutionized treatment of what was considered to be a terminal illness. Combining the two drugs is more effective than using them individually but carries a larger risk of serious adverse effects, including severe rashes, thyroid problems, and other autoimmune reactions, he says. This combination has been tried on melanoma that has metastasized to the brain, but only in a few cases and with mixed results.

Some of what Dr. Atkins shares with us is familiar by now, but much of it is not. Chemotherapy, which for many years has been the gold standard for cancer treatment, isn't effective against melanoma, one of the most aggressive cancers, he says. What's more, chemotherapy indiscriminately attacks all fast-growing cells, including healthy ones—and it causes many side effects, from hair loss to infections, neuropathies, nausea, vomiting, and fatigue. By contrast, immunotherapy drugs don't directly target cells but instead treat the patient's immune system so that it can seek out and attack the tumor cells. While there can be serious side effects with immunotherapy too, it holds great promise for treating melanoma.

Then, the magic words: Dr. Atkins invites me to join the clinical trial. It's tremendous news. Clinical trials have a limited number of patients. *I'm going to be a guinea pig or, better yet, an experimental rat,* I think to myself with a smile.

Just hours earlier, Mirek and I realized that we'd reached a wall with this whole nightmare, that there was nothing more to do but wait. Now, all of a sudden, a gate in the wall has opened and we are ready to run through it without knowing what's awaiting us on the other side. We thank Dr. Atkins and brace ourselves for the unknown.

"It can work," Dr. Atkins promises. "Absolutely, I have seen it work."

We cling to his confidence. He seems so certain.

I will have four treatments, one every three weeks, starting on April 16—just two weeks away. But first, I need to jump through a few more hoops, including a dentist's appointment, to make sure I don't have any pressing dental problems, and a series of blood tests. Most important, I need another brain MRI, to make sure I have no brain tumors besides the ones that have already been radiated. If there are any new brain tumors, I can't join the clinical trial, at least not right away.

This trial is not for patients with active, untreated brain tumors, Dr. Atkins tells me. He doesn't explain further, but later, in reading the scientific literature, I learn that it's essential that there are no active tumors, meaning ones that haven't yet been radiated. Active brain tumors subjected to immunotherapy can become inflamed, and the patient may suffer serious brain swelling, which can be deadly. At this early phase in the clinical trial, when not much is known about the response of active brain tumors to immunotherapy, it's just too dangerous to try this treatment on anyone with tumors that are still growing.

We drive home, elated and hopeful. As we pass the supermarket construction site, I realize I am desperate to see it completed. I cut a silent deal with my brain, begging it to keep any new tumors at bay so I can get the immunotherapy infusions—my best shot at survival, and perhaps my only one.

Stay clean, stay clean, I tell it. *It's our only hope.*

· · ·

A week later, a few days before the trial is to begin, I lie as still as a corpse for the most important MRI ever. I'm deeply anxious about what it might show, terrified that my last chance for life will be snatched away.

The next day, I get a phone call at work. It's the nurse from Dr. Atkins's office.

"What did the MRI show? Any new tumors? Is everything okay?"

"Yeah, it's okay," she says, her tone less excited than I think it should be. "We'll see you on April sixteenth."

I am ecstatic.

I undergo a full-body CT scan, a requirement before starting the trial, and it shows three small tumors in my lung. But we aren't alarmed. Tumors in the rest of the body are to be expected with metastatic melanoma, since melanoma cells traveling in the bloodstream often invade other organs. These lung tumors are less dangerous and easier to treat than brain tumors, and the immunotherapy will likely kill them. Even if they initially swell due to the treatment, they won't cause the same devastation as inflamed tumors in the brain, so their presence does not disqualify me from the clinical trial. Mirek and I are thrilled to get that news.

But Lance Armstrong's advice rings loudly. I decide to get a second opinion on the new MRI of my brain. I really like and feel a connection with Dr. Aizer, the radiation oncologist in Boston. So I e-mail him, tell him about our recent trip to Hawaii, mention that I'm about to enter the clinical immunotherapy trial, and ask if he'll review the MRI.

He writes back to say that he's glad I'm so physically active. "I wish more of my patients could do even a tenth of what you routinely do from an activity perspective," he says. He adds that he thinks the combination-drug immunotherapy "sounds like a great initial plan." He says he is happy to review my MRI and future scans. I FedEx him a CD of the MRI.

A couple of days later, on Wednesday, April 15, I'm at the hospital very early in the morning for a blood test, my last remaining pretrial exam. If everything is in order, as I'm sure it will be, I'll be cleared for the first immunotherapy infusion, which is scheduled for tomorrow.

At 6:22 a.m., I get an e-mail from Dr. Aizer.

Hi, Dr. Lipska, Do you have a moment to touch base over the phone today by chance? I want to check in with you. Best, Ayal

This kind of e-mail can't be good. I step outside to call him. The cherry trees are in full blossom, white clouds are rolling by in the blue sky, and it's so early that the sun is throwing long shadows across the lawn. I shiver from the cold and worry.

"Dr. Lipska, I am so sorry," he tells me. "I see new tumors in your brain. They are very small but you should get them radiated before immunotherapy."

I can't believe what he's saying.

"No, I can't, I can't wait!" I insist. "I'm going in for my first infusion tomorrow! There's no time for radiation—they'll kick me out of the trial! Dr. Atkins says I'm fine. He doesn't see anything on the scan. Are you sure?"

"The tumors are very small. They could easily be missed, but they're definitely there," he says. "One is in the frontal cortex, where it could be dangerous for your intellect and cognition, as you know so well, Dr. Lipska. You really should get them treated before you start immunotherapy."

"I can't!" I repeat. "They will kick me out of the protocol!"

For half an hour, he tries to convince me to get radiation. He repeats that the tumor in the frontal cortex could be especially problematic. Without radiation, it will almost certainly grow, and, if subjected to immunotherapy, it could also become inflamed, causing my brain to swell uncontrollably. It could quickly begin to seriously damage all the highest functions of my mind: my ability to think and remember, to express emotion, to understand language. In short, it could cut off all the things that make me human. If it causes too much swelling, it could even kill me.

"But of course, another possibility is that all the tumors will be destroyed by immunotherapy drugs, yes? Don't you think so, Dr. Aizer?" I ask.

"Perhaps," he responds, and he says again how sorry he is. I thank him and we hang up.

I stare at the darkened cell phone.

Shit. I'm dead.

I'm dead either way. If I tell anyone at Georgetown what Dr. Aizer found on the scans, they'll refuse to give me the infusion, which is my only chance at salvation. But I'm dead if I don't tell them because subjecting these new tumors to immunotherapy could kill me.

What do I do?

Dr. Atkins's report states I have no new tumors. And his nurse told me the scan was clean. They cleared me for treatment! Did he misread the MRI? Radiology is more an art than a precise science, and it's possible he didn't see them. Dr. Aizer did say they are very small.

Or maybe Dr. Aizer is wrong. Maybe what he saw aren't tumors but something else, a scar from my radiation, perhaps, or an artifact?

I don't know.

I can postpone the treatment. I can get the new tumors radiated, as Dr. Aizer insists, and then wait two weeks after radiation—as the protocol requires—before getting another scan. If that one comes out clean, then perhaps I can start immunotherapy—if there is still room for me in the trial. But if new tumors keep popping up, it will be an endless cycle: I'll get a scan that finds a new tumor, get radiation, get a new scan that finds another new tumor, and on and on and on. I can't keep radiating every single one—I'll have no brain left. Meanwhile, I'll be locked out of the clinical trial. They have very stringent timelines for being included, and no doubt there are many other desperate people eager to take my place.

This is my only shot.

I'm supposed to start the trial tomorrow.

What do I do?

The sky is so blue. What a glorious day.

No. No question. I'm doing this. It's my only hope.

I'm not going to say anything about the new tumors to anyone. I'm not telling Dr. Atkins what Dr. Aizer said, and I'm not telling Mirek or Kasia or Witek or my sister. I'm making this decision for myself, by myself. Nothing is going to stop me from entering this clinical trial. I would rather take my chances than die without trying it.

That night at home, I mention nothing to Mirek. When Kasia calls, I calmly tell her I'm looking forward to the next day. I share nothing about my dilemma or the choice I've made.

I maintain my strategic silence the next morning when I march into the hospital and head with Mirek into the infusion unit, a large room with individual patient cubicles partitioned by curtains. After I sign in and sit in my cubicle, Dr. Atkins enters with his entourage of smiling nurses and greets me.

"Are you ready?" he asks.

It's my last chance to call this off.

"So, everything is okay, yes?" I ask.

"Yes," he says.

"Will you be doing brain scans during this therapy? To check for any new tumors?"

"No, we won't need them for another three months," he says. "This is going to work."

I watch him walk away. I feel like a paratrooper jumping off a plane into the dark night, hoping that my parachute will open.

I jump.

As I sit in a reclining chair, the nurse punctures my arm for an IV that begins to drip the drugs into my bloodstream.

I rest my head on the back of the chair and close my eyes.

Maybe this will kill me. But I'll definitely die without it. Dr. Atkins believes it will work. And I trust this immunotherapy even more than I trust him.

I will live, I tell myself. *I will live.*

On the drive home, I tell Mirek my secret. "Dr. Aizer found three new tumors in my brain yesterday but I didn't tell Dr. Atkins," I say. "Nothing is keeping me from this trial."

Mirek's smile is hesitant but he nods his approval. "I understand," he says. I call Kasia and tell her too. To my surprise, she—like Mirek—approves of my decision.

"Brave mum," she says to me in Polish.

A few days later, Kasia and I have a phone conference with Dr. Aizer. He reiterates that it could be dangerous for me to continue with immunotherapy with three new tumors growing in my brain. When I tell him I will not be getting any brain scans for three months, he becomes even more worried. But Kasia and I remain united around my decision. We listen but we really don't want to hear his concerns. I don't know it at the time, but after we hang up, Dr. Aizer goes into my sister's office at the Brigham and tells her he is very worried about me. She hears him out but she knows I have made up my mind. She keeps the conversation to herself until long after I have completed my treatment.

On May 5, three weeks after my first infusion, I go in for my second. Mirek and I awaken early and drive to the Georgetown hospital to try to find a precious parking spot in the underground garage of the old, cramped complex. We proceed through a maze of corridors to the Lombardi cancer center, passing by a portrait of the pope that the hospital staff use as a landmark for giving directions. ("Go straight by the pope to get to the infusion center," they say, or "Turn right by the pope if you're here for an MRI.")

As usual, the waiting room at Lombardi is filled with patients, some bald from chemotherapy, some in wheelchairs, some limping on their canes. But most look healthy and normal. The techs draw my blood and we wait for the lab results. After a couple of hours, we see the doctor, who evaluates the blood tests to decide whether I am well enough to receive an infusion that day. It feels like waiting to learn if I've won the lottery. As before, I obsess that I will be rejected because of an abnormal blood result or some other lurking danger.

But it doesn't happen. Physically strong and optimistic against all odds, I go through the second cycle of immunotherapy without major problems. I am now halfway through the twelve weeks of the clinical trial and feeling good. With each IV drip, I imagine my newly emboldened T cells leading the army of my immune system to attack and defeat every melanoma cell in my body. I will it to happen, I will all of the cancer cells to die. They *must* die.

I'm full of hope and energy. Nearly every day, I run or walk a few miles. Nearly every day, I go to work and complete my tasks with no problem. With every bone in my body, every neuron in my brain, I believe that I am on the mend.

And then—everything breaks down.

I never see it coming.

4

Derailed

*S*ometime after the second infusion, my body turns on me.

Modified by the drugs to sense danger everywhere, my immune system has been on high alert since the first infusion. Now, after the second one, it begins to attack not just the tumors in my brain but healthy tissue throughout my body. This autoimmune reaction causes inflammation in my skin, thyroid glands, and pituitary gland, a tiny structure in the innermost part of the brain that controls the flow of hormones to the body's other glands, including the adrenals. Soon, my thyroid stops functioning, forcing me to take replacement thyroid hormones; I also begin taking prednisone to stop the rashes and fill in for the natural steroids that my adrenal glands have stopped producing and without which I would experience serious fatigue, muscle weakness, and weight loss.

My skin bothers me most. From my scalp to my feet, and especially on my back and butt, I'm covered in a red, itchy rash. I have trouble sleeping, and I have to scratch, scratch, scratch. I lather soothing steroid creams all over my body, which help for a while, but soon the itching comes back, and I scratch again. The only relief I get is when I stand in the shower with lukewarm water cascading over me.

And there's another side effect that I can't ignore anymore.

"I really need to take care of my arm," I say to Mirek. "Look how swollen it is. It's very uncomfortable."

When my left breast was removed six years ago, almost all of the lymph nodes under my left arm were taken out too. Without the nodes, lymph fluid can't drain properly, and it builds up in the tissues in my arm, causing it to swell, a condition called lymphedema. My engorged arm has been a nagging reminder that I'm not 100 percent healthy, so over the past few years I've ignored it and endured the discomfort and puffiness. Now, however, immunotherapy is aggravating the lymphedema, a side effect that I knew was likely. While I see it as a fairly benign consequence of the treatment that may save my life, it really hurts. I can no longer put off getting help.

I call the reception desk in the department of physical therapy at nearby Inova Fairfax Hospital and ask for an appointment. There's nothing open until mid-June, weeks from now. I'm surprised and dismayed by the delay and try to convince myself the time will fly by so fast I won't notice. But my arm is really sore.

I decide to take my mind off it by making a quick trip to see my daughter and her family in New Haven. It's been a month since we were together and I yearn to see them, to spend as much time with them as I can while I still have the time to spend. My third infusion is scheduled for May 26, a week away. I'll be on a northbound train the very next day.

May 27 dawns hot and humid, a taste of the brutal summer that will soon stifle the mid-Atlantic. My left arm is swollen and aching, and my full-body rash is driving me crazy. But my physical discomfort is dwarfed by the joy I feel at the thought of seeing my daughter, son-in-law, and grandsons. Not for a moment do I consider canceling my trip.

At noon, Mirek drops me off at Union Station in downtown Washington, DC, and I board the Amtrak to New Haven. I clamber onto the train with my few belongings in a small suitcase and head for the quiet car, where cell phones and loud conversations are banned. I find a window seat in a row with no other passengers, nestle into the cushions, pull a book from my purse, and cherish the solitude.

The train rumbles slowly across Maryland, then New Jersey. Then, in the middle of nowhere, it grinds to a halt. Through the window I see empty fields, green and expansive pastures, and a few trees dotting the landscape. There's no station in the vicinity, not so much as a house.

After a moment, the lights in the train and the air-conditioning system go out. All of the electricity has been shut off.

We wait in total silence, a quiet that I appreciated a short time earlier. But not anymore. This silence is irritating. It lacks purpose.

I place my bloated arm on the narrow windowsill. It's too high and makes my arm even more uncomfortable. But the armrest doesn't work either; it's too low. My arm is painful, my hand swollen. I stare at its fingers and palm, so fat and tender that they look like they could burst.

Why didn't I call the therapist sooner?

I try to focus on my book, try to be patient and relax, to no avail. My discomfort persists and the delay grows; minutes tick by and we're still not moving. There are no announcements, and no one in the train car seems to know what's going on. Finally, after at least half an hour, the loudspeaker crackles.

"There's a problem on the tracks. A tree fell," a voice announces. "We're waiting for a maintenance crew to arrive and remove it. Then we'll be on our way."

More time goes by, and still nothing happens. The car is hot, and I'm thirsty. My skin feels like it's on fire. In addition to the pulsating pain in my arm, I realize that I also have a headache—a mild but nagging throb that fills my whole skull.

Two hours pass before we begin to move. But even then, we dawdle along at a pace that seems even slower than before. It's nearly as excruciating as if we weren't moving at all.

I storm out of the quiet car into the vestibule and call Kasia, fuming.

"Unbelievable! This Amtrak is good for nothing!" I hiss in Polish. "How can they keep people waiting, uninformed, left to fend for themselves with no food or water? Complete lack of responsibility!"

Kasia listens patiently and tells me she can't wait to see me. Her voice, so dear, does nothing to calm me.

It takes seven hours to get to New Haven instead of the usual five. When the train pulls into the station, I loudly share my unhappiness with everyone around. "Even five hours would be too long!" I say, daring anyone to challenge me. "The infrastructure in our country is pitiful. In Europe, this trip would take a fraction of the time." I'm tired and hot, and this headache won't go away.

I hail a cab at the station; in fifteen minutes, it pulls up to Kasia and Jake's home.

When I walk in the front door, Lucian and Sebastian leap at me with such force that I almost end up on the floor. "Babcia! Babcia!" they scream in unison. "I love you, I love you so much! I missed you!" I kiss their ketchup-stained faces and hug them and don't want to let go.

Kasia runs out of the kitchen to greet me. "Mama!" she exclaims. "I'm so happy you're here!"

She kisses me, and I press my body against hers with all my might. I want to feel the warmth of my daughter, to let her know that I've missed her and am very happy to be with her. From a beautiful little girl, she's turned into a gorgeous, mature woman, smart and exceptionally devoted to her family and her challenging work. I want to tell her, as I have so many times before, how proud I am, how pleased to see her so pretty and so accomplished.

But that's not what I say.

"Amtrak *sucks!*" Those are the first words from my mouth.

She looks a little shocked.

"I can't tell you how long that train ride was," I say forcefully. "I will *never* take that train again!"

"Mum, come in and sit down. Let's relax and—"

"It's irresponsible to keep people on the train for so long. It was horrible."

I see her staring at me, imploring me to let it go, but I have no intention of doing that. I've been wronged, and I want her sympathy. "There is no excuse," I continue. "It's a shame that in this rich and technologically advanced country, the trains are in such poor shape. In Europe, they run so much faster. Can you believe how long I was on the train?"

Sebastian and Lucian tug at my hand, trying to pull me into a game. But I want the boys to understand what I've just been through too. It was a *terrible* experience.

"Amtrak sucks!" I say it again. And again. Lucian and Sebastian are getting bored with my tirade, and soon they disappear into their room to continue their wild games, screaming and laughing.

"Okay, Mum, enough about the train," Kasia breaks in. "You're here now. What can I get you? Do you want to lie down?"

Enough? I think. *I am deeply aggrieved!* "The train was horrible—"

"Let's talk about something else," she says gently.

"Why can't I express my opinion?" I shoot back bitterly.

Kasia tries to brush off my outburst. She tends to the boys and starts preparing our dinner. But I can't move on. I'm irritated by Kasia. I'm irritated by the boys, by everything. Suddenly, I am so, so tired. And this headache—it's not going away.

I stay in New Haven for two days as planned. But my time there is not nearly as restorative as it was intended to be, either for me or for my family.

I can't stop talking about my train ride. I bring it up with Kasia and Jake—and with their friends, too, when they drop by to say hello and wish me well. They listen politely but I can see from their expressions that they're thinking, *Why are you telling this story? Why is it such a big deal?*

But it *is* a big deal. It's a huge deal. If they can't see that, there must be something wrong with them.

Amtrak sucks! The refrain circles around in my mind like a toy train looping a closed track. *Amtrak sucks!* I say it out loud, too, over and over, to anyone who will listen.

It's not only Amtrak that is the object of my ire. I'm irritated if our lunch is even five minutes later than Kasia promised. I can't stand how loud the boys are. I find everything my family does annoying, and I tell them so.

On the second afternoon of my visit, Sebastian comes running over, laughing loudly, and bumps into me. It pisses me off. "Be quiet!" I tell him. "Just stop it! Stop it!"

He looks like he's about to cry. "You're so mean!" he says.

"Oh, come on! You can't be that sensitive! Can't you take criticism? That's just crazy!"

He bursts into tears and runs from the room. Kasia comes in from the kitchen.

"Really, Mom," she says. "You are being mean. It's not like you."

I cannot believe what I'm hearing.

She's siding with him? I'm *mean? Is she serious?*

I turn away. I don't want to talk to any of them. I go into the guest room and close the door.

Why is Kasia arguing with me? I wonder as I lie in the darkened bedroom nursing my swollen arm. *I deserve better treatment than this.*

I'm not the only one who's baffled by how this visit is going. I will learn much later that, as I'm sequestered upstairs on this late spring day, Jake and Kasia, downstairs in the kitchen, are talking about me, quietly so I can't hear them. They're both surprised by the way I snapped at Sebastian, to whom I'm always so loving. While I never fail to speak my mind, I'm also reflexively warm and affectionate toward my family. They now find me distant and anxious, and my obsession with the train puzzles them. They can't figure out what's going on.

Kasia thinks it must be the anxiety of the experimental treatment and the horror of facing my own mortality. Maybe I'm depressed, she speculates. But Jake isn't so sure. I've had brushes with death before, he notes, but I was always open and vulnerable and shared with them my fears and emotions. This feels off, they agree.

It may be obvious to them that I'm acting strangely but I can't see that my behavior is unusual. Nor can I see the confusion and suffering it is causing. Upstairs in the guest room, I'm in my own world, fixated on their poor treatment of me and the gross incompetence of the American railway system.

What's wrong with all of them? Kasia is not being as warm as she usually is. The boys are too noisy—they're getting spoiled. Amtrak sucks!

My headache is back. This damned heat.

Compared to the itching and other side effects of immunotherapy, the pain in my head feels like a minor inconvenience. Still, I called the nurse at Georgetown yesterday, just to be sure. But when I described it as mild and intermittent, we decided it wasn't a big worry, although she asked me to keep an eye on it. It's certainly not the kind of severe or sudden headache that would set off alarm bells for me, Kasia, or my doctors. I've soldiered through much worse, I think—and I fail to see it as a warning sign.

• • •

I don't realize it, and no one around me does either, but deep inside my brain, a full-scale war has erupted. The tumors that were radiated are shedding dead cells and creating waste and necrotic—or dead—tissue. These old tumors are also under attack by immunotherapy, as are the three new tumors that Dr. Aizer found shortly before I was to enter the clinical trial. Mortally wounded by my modified T cells, the cancer cells from the six tumors found between January and April are like tiny dead bodies. They must be broken up into smaller particles and removed from my brain through the blood and lymphatic systems. Throughout my brain, the tissues are inflamed and swollen from the metastases and the double assault of radiation and immunotherapy. What's more, my blood-brain barrier—which normally prevents circulating toxins and other substances from entering the brain—has become disrupted by immunotherapy and is leaking fluids through small vessels and capillaries. The fluids are pooling in my brain, irritating the brain tissue and causing it to swell, a condition called vasogenic edema.

All of this is wreaking havoc on my brain, just as my behavior is wreaking havoc on my family. Although I knew I could pay a heavy price for the chance to live, I had no idea how high the cost would be. My brain—in particular my frontal lobe, which Dr. Aizer was especially worried about because it controls higher cognitive functions—is a deadly battlefield.

And my life is in serious jeopardy. Composed of hard bones, the skull is not flexible; it can't expand outward to release pressure in the brain. When the brain swells, there's only one place it can go: the foramen magnum, the hole in the base of the skull through which the brain stem exits into the spinal cord. The most primitive part of the brain, the brain stem controls primal functions including respiration, heart rate, and blood pressure. If the brain stem is squeezed by swelling or is otherwise injured, a person can go into cardiopulmonary arrest—the heart and breathing stop—and die.

If I could have recognized that my frontal lobe was under attack and the effect it was having on my personality, I would perhaps have seen some parallels with the famous case of Phineas Gage, a railroad worker who suffered a horrific injury in the mid-nineteenth century. Gage's personal tragedy marked a turning point in the study of the brain. He had been using a long iron rod to pound blasting powder into a mass of rock when the explosive suddenly went off, sending the rod shooting through his head like a javelin. It entered his left cheek and passed through the left side of his brain, obliterating much of his frontal lobe, then exited through the top of his skull and landed some eighty feet from where Gage stood. Incredibly, the twenty-five-year-old survived, living for another eleven years with a huge hole in his head—and with dramatic changes in his personality. Once a likable guy, he began to swear constantly, couldn't follow through on basic tasks, and seemed to care about no one but himself. His behavior grew so bad that he was fired, after which he lived a peripatetic existence, ultimately perishing after a series of convulsions that may or may not have been connected to his devastating injury.

Gage's misfortune taught us something critical about the connection between the frontal lobe and the mind—although not the lesson that was assumed at the time. Contemporary scientists theorized that the portions of Gage's brain that were destroyed in the accident were responsible for controlling his personality, but we now know the truth is more complicated. Emotions, which form the foundations of our personalities, are not contained in a single brain region, as once believed, but rather are distributed throughout the brain in a complex network that we don't yet fully understand.

It is clear, nevertheless, that the frontal lobe is intricately connected to the ways in which personality is expressed. People with damage to their frontal lobes—whether as a result of head trauma, like Gage; cancer, like me; or a neurodegenerative disease, as with Alzheimer's patients—often undergo significant personality changes. In some cases, these changes are truly bizarre, combining noticeable disinhibition with little appreciation or concern about the consequences of one's actions. More extreme examples may include loud and frequent swearing or sexually inappropriate behaviors.

Most mental conditions—from Alzheimer's to schizophrenia, from bipolar disorder to depression—involve some sort of change to a person's emotions and thus his or her personality. But whenever someone exhibits significant personality changes, especially over a relatively short period, it's also possible that a frontal-lobe problem —a tumor or injury, for example—is to blame.

Like my headaches, the changes in my personality signal that something serious is going on. Squashed like Jell-O in a jar and pushed out of place from the swelling, my frontal cortex can't perform its supervisory function of telling me to stop and think before I jump into action. In a sense, this crucial part of my brain has reverted to an earlier state, not unlike the brain of a small child who hasn't yet learned how to exercise self-control or navigate delicate social situations.

I have no idea any of this is happening. If I notice anything amiss, I just assume I'm stressed out—from the heat, from the exertion of the trip, from the noise and activity of life with my grandsons. All I need is to get back to my own house and my regular schedule, which is much less hectic than theirs. I long for peace and quiet. I miss Mirek, and I can't wait to be home with him.

I leave New Haven on May 29, the day after I erupted at Sebastian. My shaken daughter and grandchildren escort me to the train station. As I kiss them goodbye, I know I will miss them, yet I'm eager to get home.

The journey back is uneventful, and Mirek meets me at Union Station. From a distance, I easily spot his car, a green Volkswagen Passat outfitted with a roof rack for our bikes.

As I step off the train, he is beaming. "I'm so happy to see you," he says, leaning in for a kiss. "I missed you."

I don't kiss him back. "I'm very tired," I snap. "I want to go home."

He gives me a puzzled look tinged with hurt. "Did anything bad happen?" he asks. "I am sure you had a wonderful time, no?"

"Why are you asking me these questions now? I'm tired!"

He retreats into silence, but I pursue him. "You always ask me so many questions," I hiss. "What's wrong with you?"

His eyes are glistening. Are those tears? I don't care.

Mirek says nothing more. We drive home in total silence.

5

Poisoned

*A*s June arrives, I return to the routine of what's become normal life for me: a never-ending parade of doctors and medical appointments even as I continue to work full-time. At the office, I find the minor shortcomings of my employees very irritating. But instead of letting these small things slide, as I normally would, I begin to criticize them frequently.

Of course I find things irritating, I tell myself. *I'm tired of being sick. I'm tired of my rash, my swollen arm. I'm tired of everything.* And my headache continues to come and go.

When the date finally arrives for the physical therapy appointment for my lymphedema, I don't feel like going. Although my arm is still uncomfortable, I loathe the idea of another hospital, another treatment. These medical visits are relentless reminders that I'm ill.

And this visit is especially irksome now, when I'm trying to feel hopeful. My melanoma will retreat in the face of the magnificent, novel treatment I'm receiving. I *know* it.

But I'm a woman of my word, so rather than cancel at the last minute, I keep the appointment and go. It's a short drive along back roads to the hospital from my home; afterward, I'll head directly to work.

I know our local hospital well. I've been here many times for the various minor surgeries that Mirek, Witek, and I have endured over the past thirty years. But today, as I pull into the entrance to the parking complex, I wonder if I'm in the right place.

Everything looks totally unfamiliar. I don't remember the parking area having this layout.

Did they change it?

I drive into the large, multilevel garage. There are no spaces on the first floor so I continue skyward. I drive up, up, looping for what seems like forever, in circle upon circle onto higher and higher levels —but still, I find no empty parking spots.

I emerge at the very top of the garage, where I'm momentarily dazzled by the sunshine. "In this heat, the car will be unbearable when I return," I say to myself as I park.

I take the stairs down, down, down to the first floor of the garage. But once I'm there, I can't locate the hospital entrance.

Did this change too?

I wander about for a minute and finally discover the front door, but once inside, I find myself in a confusing maze of long corridors leading in all directions, lined with doors opening to who knows where.

I'm lost again. *Have they changed everything in this place?*

Irritation wells inside me. "Why did I have to come here? This is so stupid," I grumble. "Where's the office? Why don't they make it easier for patients to find their way around?"

I ask several people for directions but no matter how much they try to help, I cannot find the physical therapy department.

I can't believe they're doing this to me! I'm sick—how could they put me through this?

Somehow, I finally stumble upon the front desk of the PT department. I am seething.

After checking in, I take a seat in the waiting area—but any relief I feel at having found the office quickly dissolves. On the couch across from me, a little boy is coughing and crying. He nags his dad to take him out of the room.

I stare at the child with annoyance. *Why on earth would they let a sick kid into this room? I'm very ill. I can't afford to be around someone like him!*

As he continues to cry, my loathing increases.

Isn't physical therapy for adults? Sick children should have a separate clinic. They should be isolated! He's going to infect me!

I hate the little boy. I hate his father. I hate this place.

This torment continues for a long while until, at last, a woman in hospital scrubs enters the waiting room and calls my name. "I'm Theresa," she says with a smile. "It's nice to meet you."

Such a strained, dishonest smile. So insincere. What is she up to? I'd better keep a close eye on her.

She leads me into an examination room and offers me a seat, then begins to inspect my arm.

"The lymphedema is really advanced," she says. "You've waited way too long. The swelling may be permanent. I'm going to explain how we should treat it so that it doesn't get worse, but you have to follow my instructions very carefully. If you don't, it could be dangerous to your health. Your arm will be prone to infections."

Why is she droning on and on? This is such a dreary, boring, awful place.

I begin to wonder what we'll have for dinner tonight. *Did Mirek get the salmon from the grocery store? I'll bet he forgot. He's always forgetting everything I ask him to do. How could he—*

Her voice momentarily interrupts my thoughts. "Let me show you how to bandage the arm," she says. "You'll wear this bandage for the next month or two. It's very important—do you understand?"

What time is it? I need to get home. Especially if Mirek forgot to shop. I need to get dinner ready.

She eyes me. "You *really* need to do this," she says firmly.

I pretend to listen.

"After you're done with the bandage, you'll be using a compression sleeve like this," she says, holding out a long, flesh-colored tube designed to span the length of an arm from knuckles to armpit. "You will have to wear another sleeve at night to keep your arm compressed and prevent lymph fluid from pooling."

I glance at the sleeve. It's ugly and silly.

"Are you kidding?" I scoff. "Do you really expect me to wear that ridiculous thing? It looks like a medieval torture device."

She doesn't respond.

Who does she think she is, sitting there so smugly? "I'm a professional woman with great responsibilities," I continue. "How on earth would I look wearing these ludicrous bandages and sleeves? They may be good enough for someone sitting at home all day but not for me. I work in a serious place. I supervise a large department. You must have something better than that."

She keeps watching me, silent.

I know better than she does. "Why don't you just massage my arm and let's be done with it?" I propose.

"Massage will only work in combination with these compression sleeves," she says. "This is a serious condition. It needs immediate attention and ongoing treatment."

I don't like her expression. She's arrogant. I knew it the minute I saw that fake smile. "I am not going to wear anything on my arm," I say. "Forget it."

"You need a series of regular visits," she insists. "And you have to stop arguing with me."

"A series of visits?" I start to laugh. "I have no time for such *bullshit!*"

I stand up and give her a withering look, then wheel around and stomp out the door, through the waiting room, and into the hallway. "What kind of nonsense was that?" I say aloud as I leave.

What a waste of time this was! I'm never coming back. Appalling! They have absolutely no idea what they're doing.

I find the stairs in the parking garage and march directly to the highest floor, all the way into the sun. I get into my car and accelerate down and out of the parking garage in a long, spiraling swoop. Finally, I can head to work. I'm determined to put this absurdity behind me and get on with my day.

By this time, the highway is no longer congested by rush-hour traffic.

Of course, there are no cars on the highway anymore—everyone's at work! I would be, too, if I hadn't spent well over an hour in that stupid hospital.

It's an easy drive on the Beltway to the sprawling NIMH campus in Bethesda. This is the largest biomedical institution in the world; almost twenty-one thousand federal employees work in dozens of buildings situated on hundreds of acres of a former private estate.

Even though I'm exhausted by that pointless physical therapy appointment, I put in a long day at work, supervising all aspects of the brain bank. I'm bombarded with questions as soon as I arrive. One of the technicians asks about a potential brain and whether we should accept it; as soon as he leaves, another technician comes in with a similar query. After she leaves, I respond to a dozen e-mails from researchers around the country eager to receive some of our brain samples, and then I review the most recent data about the tissue samples we are storing.

Each time I get up and head into the lab to check on my employees, I pass by a bowl of chocolates on my assistant's desk. She always has candy out, and I always avoid it. I don't like to eat unhealthily, especially sweets. But yesterday, the chocolates looked so good that I ate them throughout the day. I couldn't stop. Today, it's the same; every time I walk by, I grab another one and pop it in my mouth. Sweets have never tasted so irresistible.

Early one evening a few days after the physical therapy appointment, I'm in the kitchen, chopping vegetables and slicing meat to make a stir-fry for our dinner. I'm sipping a glass of wine, trying to relax, when I hear a knock on our front door. Mirek is upstairs working in his office so I answer it.

A man about thirty years old with a broad smile is standing on the front step.

"Hello, Mrs. Lipska!" he says brightly.

How weird—he acts like he knows me! I've never seen this man before. What does he want? Something is very wrong—I can sense it. Something dangerous.

Without waiting for my invitation, he steps forward as if to enter our house.

I block the door.

"I'm John," he says. "From pest control." He holds out his hand. I don't shake it.

"*Who?*" I demand.

"John. We provide pest-control services for you, remember?"

He's definitely up to something.

"We've been providing termite inspection service to you for over twenty years," he says, speaking more slowly now.

That change in his voice—he knows I'm onto him.

"This is our scheduled visit," he continues. "May I do the inspection?"

"The inspection? Oh, really?" I make sure my sarcasm is clear. "Why are you here today, exactly?"

He gives me a puzzled look.

"What," I ask again, "do you think you're going to do?"

He starts to talk about termites. That reminds me of something urgent.

"Ants!" I shout. "They're everywhere!" I dash into the kitchen. "Come! Look here and here!"

I point at the windowsill, where a few tiny ants are marching along the wall to the back door that opens onto the outside deck. "Ants! See? And you absolutely have to see the stain on the wall in the basement. It might be mold." My words pour out in a torrent. "Go look, quickly!"

He rushes downstairs to the basement. I'm relieved to be rid of him but a few minutes later he returns, talking about something or other. The only word I hear is *chemicals*.

He's going to spray some chemicals.

"Chemicals!" I jump as if someone's poked me. "What do you mean, *chemicals*?"

He looks scared.

I knew it! I've caught him.

"Our chemicals are very effective against ants and fungus," he says, but his speech is halting and uncertain.

Aha! His little game is up.

"We have another spray, for the termites." He pauses and then adds, "Don't worry. They're all very safe."

"Safe? Chemicals?" I shout. "Chemicals are poison! Don't you know that? How can you say that they are safe?"

"Well, customer safety is key to our—"

"Then tell me, what's in these chemicals?" I demand. "What compounds do you use?"

He stares at me blankly.

I have him cornered! "You have no idea, do you? Safe? Ha! I'm a chemist! You can't fool me. I have young grandsons! Are you trying to poison them? To poison us all? Is that your plan? All chemicals are toxic. I forbid you to use any chemicals in this house."

Somebody approaches behind me, and I realize that Mirek has come down the stairs.

"Hello, how are you?" Mirek says to the young man.

Why is Mirek greeting him so cordially? This stranger is trying to poison us!

Mirek turns to me. "Don't worry, he won't do anything today," he says soothingly. "He'll just do the inspection. Here, I'll sign the papers." Mirek turns to some documents the young man has placed on the kitchen island.

"No way!" I shout, inserting myself between him and the island. I lean in toward the young man and yell, "You are fired!"

His face is frozen in disbelief. Before Mirek can say anything, I continue. "Not only are you not working for us anymore, but I'm going to call your manager and tell him that you are *totally* incompetent. How can you not know the chemical composition of your own spray!"

Unbelievable! What an idiot!

I turn and storm away, leaving Mirek and the stranger alone in the kitchen.

These kinds of changes in normal behavior often signal that something serious is occurring inside a person's brain. My emotional overreactions—anger, suspicion, impatience—suggest that my frontal lobe is undergoing catastrophic changes. But these warning signs are lost on me. As an expert on mental illness, I, more than most people, should be able to see that I'm acting strangely. But I can't. Although I don't know it yet, my six tumors and the swelling around them are shutting down the frontal cortex, the part of the brain that allows for self-reflection. Paradoxically, I need my frontal cortex in order to understand that mine has gone missing.

This inability to recognize my own impairment is often observed in people with mental disorders. Known as anosognosia, or lack of insight, it's a feature of many neurological and psychiatric conditions. Little is known about which brain regions are responsible for lack of insight, but some investigators suggest that it may be related to dysfunctions in the midline of the brain, which separates the right and left hemispheres. Damage to the right hemisphere may also be involved.

In schizophrenia and bipolar disorder, a lack of insight into one's condition is thought to be a manifestation of the illness itself rather than denial or a coping mechanism, as it may initially seem to be. About 50 percent of people with schizophrenia and 40 percent with bipolar disorder cannot understand that they are sick, so they have no real awareness of their condition and won't accept their diagnoses. If they experience hallucinations or delusions, they don't see them as a sign that something is wrong with their brains; even the most dramatic symptoms, such as hearing voices or believing that they are God, are indistinguishable from reality. Because people with schizophrenia and bipolar disorder who suffer from a lack of insight don't believe they are ill, they're often very resistant to psychiatric treatment. They may not take prescribed medications or participate in behavioral therapies. And there's no cure, at present, for this kind of lack of insight.

Just like someone with schizophrenia, I don't think that there's something seriously wrong with *me*. I think I'm absolutely fine mentally. If anything, I believe I'm just stressed or tired—worn out by the poor design of a medical facility, by the inexcusable wailing of a child in a hospital waiting room, by the appearance of a strange, pushy man at my front door. I do not connect these dots or deduce that the problem is in my head rather than someplace beyond it. I have no reason to think that my reactions to these incidents might be related to my tumors or cancer treatment, nor does anyone around me—at this point, I am not getting any MRIs, which would reveal what is happening inside my brain.

And so, as my confusion grows, my brain fills in the gaps between what's in my head and what's happening around me with conspiracy theories. I become increasingly suspicious of my family and my col-

leagues at work and increasingly dissatisfied by everyone's perfor-
mance of even simple tasks. I'm certain that people, especially the
members of my family, are plotting against me.

Kasia doesn't really like me anymore. I don't think Mirek does either.
Why are they talking about me? I can tell they're hiding things from me.
But what? What are they hiding?

Feelings of suspicion—sometimes rising to the level of paranoia
—can be a symptom of many types of mental illness, including Alzheimer's disease. Alzheimer's patients may accuse their romantic
partners of cheating on them or their caretakers of stealing property or trying to harm or even kill them. While neuroscientists
don't really understand the networks or parts of the brain related to
paranoia, in some cases this condition is attributed to temporal lobe
damage.

And while the turmoil in my brain may be responsible for my behavioral overreactions, it's true that my feelings are not completely
irrational. I have good reason to be suspicious; after all, my worried
family *is* talking about the way I'm acting. To their dismay, all of my
least likable characteristics—my need for organization, my insistence on doing things my way—are growing more exaggerated as the
days go by. I'm becoming the worst version of myself: selfish and un-
concerned about other people's feelings. I've lost empathy, the trait
that was always so strong in me. Whereas once I listened patiently to
Kasia on the phone as she described her workday or challenges with
the kids, I now cut her off. I am losing my emotional connection to
the people closest to me, especially my caring husband.

Why are some people highly empathic while others are profoundly selfish? Like so much about human behavior, we don't know
for sure. Empathy, like other complex behaviors, is not situated in
one isolated part of the brain but regulated by a wide network of connections among many brain regions. An intricate mix of genetic and
environmental factors are probably at work: how each brain is structured and internally connected, how a person is brought up, and

where and in what culture he or she is raised. Each individual's personality is a result of complex interactions among countless factors that influence the brain's function.

However, some scientists think that some brain regions may be more involved in empathy than others, and these include the frontal cortex, the temporal lobe, and the insula, a cortical region located deep inside the brain between the frontal and temporal lobes. If that's true, it may explain why loss of empathy is often a core symptom of a kind of dementia called frontotemporal dementia (FTD), which is caused by a progressive and ultimately fatal neurodegenerative disease.

Dementia is a broad term that refers to certain mental declines, such as loss of memory and social and cognitive abilities, that are serious enough to interfere with daily life and that have lasted at least twelve months. The most common cause of dementia is Alzheimer's disease, which accounts for 60 to 80 percent of all dementia cases and is characterized by lapses in memory, language, or executive functioning. Certain other neurodegenerative diseases also cause dementia, and so can stroke, traumatic brain injury, and infections such as syphilis and HIV. The World Health Organization estimates that roughly forty-seven million people worldwide suffer from dementia of some kind, with nearly ten million new cases diagnosed each year.

Since my symptoms are so new and transient, I don't come close to meeting the criteria for dementia. But some of the personality changes I've begun to experience during the trip to New Haven are similar to those seen in cases of frontotemporal dementia, which, as its name suggests, affects the frontal and temporal lobes. FTD typically strikes people at a younger age than Alzheimer's, with 60 percent of cases occurring in people forty-five to sixty-four years old; that is, in middle age. Because the frontal lobe is involved, patients often become disinhibited and lose their judgment, and it is sadly apt that FTD is sometimes called the midlife-crisis disease. Some people become sexually inappropriate; some go on wild shopping sprees, become financially irresponsible, or eat junk food with abandon. They may behave as if their ids are running amok with no superegos to override their impulses and desires. People with FTD typically lack empathy, and also are convinced that they're doing nothing wrong. This lack of insight is a core criterion for FTD and for many other mental disorders, including schizophrenia—the disease I have spent so much of my life studying.

While I don't have frontotemporal dementia or schizophrenia, the swelling in my brain is causing me to act like someone with mental illness: I am here in body but not always in mind. The people around me recognize me yet don't. They are struggling to understand why I could possibly be behaving so strangely. And I am oblivious to their concerns.

The world around me seems more and more peculiar, and my confusion often morphs into anger.

Everything that everyone does is so irritating. More than irritating—infuriating!

What's the matter with everyone at work? Why can't they do things the right way? Why is it always up to me to correct their mistakes? Mirek isn't any better. Everything he does is wrong. And no matter how much I point it out, he keeps screwing things up. It's unbelievable.

My complaints are relentless. "Why did you put a napkin here and not there? It makes no sense!" I say to Mirek as I'm preparing dinner. Or, "Why are you still sitting? Don't you see that I need your help *right now?*"

Each time I snap at him, he gently asks me to calm down. I hate that—it's so stupid and weak. It just makes me angrier.

Why is Mirek such a wimp? What happened to him?

He frets over my health, always asking me if I need anything, urging me to do the things I enjoy—to go for a run or take a bike ride. It irritates me. More and more, I avoid his eyes. I don't care how it affects him. I don't care what he's thinking or feeling. I don't care what he is going through at work or anywhere else. I have more important things to focus on.

What will I have for breakfast? Is the table setting complete? And now Mirek has put the forks somewhere that I can't find them! Why would he do this to me? Where is the salt? I can't remember what I was planning for dinner. For the life of me, I just cannot recall. It really bothers me. And where is Mirek?

Bothered by my short temper and egotism, my family tiptoes around me. And, out of earshot, they quietly share their concerns. Upstairs in his office, on one occasion that I will learn about much later, Mirek speaks with Kasia on the phone, telling her that I'm being difficult, so difficult that he is really struggling. She can tell that he's trying hard not to cry.

I'm not the same woman they've always known, they agree. I'm an angry, overly critical, selfish version of myself. My characteristics are basically the same, for the most part, but exaggerated. I am an over-the-top caricature of myself.

But my behavior is not so bizarre that it sends up red flags about my health. I've always spoken my mind, more so than anyone in the family; they're used to that. And my concern about the chemicals in the pesticides, for example, wasn't unreasonable, they admit; chemicals can be dangerous, after all, so the fact I ripped into the pest-control guy wasn't completely out of line.

So my awful behavior continues unchecked. And for my part, I remain unaware that anything is amiss. With a brain that's not functioning properly, I am singularly focused on my own needs and entirely blind to the signals that something is seriously wrong with me.

There is one thing I care about more than anything else: getting that fourth and final infusion. I'm going to finish this treatment even if I have to drive myself to the hospital. Even if I have to walk the twenty miles to get there, crawl into the infusion unit, and stick the IV into my own vein. I'll do it. I'll do whatever it takes.

6

Lost

*A*t *the office, I work* long hours just as I did before my diagnosis. I act like nothing is different. I review scientific articles and manage my large staff, making detailed plans for the institute's ever-expanding brain bank. We continue collecting postmortem brains and setting up scientific collaborations with colleagues across the country at a faster and faster pace to support increased demand as more people in the scientific community learn about our brain bank. I assure my supervisors that I'm back to normal, and I send e-mails with cheery subject lines like *I am feeling good!*

And I *am* feeling good! I'm staying optimistic about my prospect of surviving this deadly cancer. While I'm no longer as strong as I was before I began immunotherapy, I'm still capable of powering through a normal workday—and, when the occasion calls for it, summoning great bursts of energy for a project or a meeting. I believe I'm doing very well, apart from the tumors in my brain.

But of course, I'm not.

Increasingly, I struggle with some tasks, and I'm having trouble focusing on what I'm doing. Reading is especially confusing. I begin delegating some of my work to my employees and sending e-mails in all caps—the electronic version of shouting, something I've never done before. On one occasion, instead of proofreading an article for a prominent academic journal myself, as I've always done, I immediately forward it via e-mail to a postdoc with a blunt note: PLEASE DO THIS. Another time, I e-mail the organizers of a professional conference whom I'd asked to make hotel reservations for me:

> Thanks. These are super special cicrumstances for me, i am batting a deadly disease. As a federal employee, i have to wait for travel approval and can only use gov rate fee h
> For hotel. I tried to ask for accommodation a few weeks ago biy to mo avail. Please help! Thanks. Barbarag

I see nothing amiss with this e-mail, and no one says anything to me about it.

Nor do I recognize that I'm becoming more and more uncaring about what other people think, and more disinhibited. At some point in June, for instance, I stop pulling down the blinds in the bathroom window at home when I'm showering. I just stop caring about who might see me. It's just too much work—and why would I block a nice view into the park?

It's around this time, in June, that I go for a run through the neighborhood without my prosthetic breast and with hair dye dripping all over me, surprising Mirek with my bizarre appearance when I return home. I see nothing off about the way I look.

I don't realize what's happening at the time, but this lack of inhibition and judgment are common in people with frontal-lobe problems due to dementia, stroke, injury, swelling in the brain, or any number

of issues. The frontal lobes give us the capacity to predict the consequences of our behavior and avoid actions with expected adverse reactions. Each of us makes thousands of judgment calls every day, in
most cases without even having to think about them. When a person
suddenly breaks normal social rules, as I'm doing now, it's a strong
indication that the frontal lobe is not working properly.

Without a functional frontal lobe, my brain is like a horse gallop-
ing dangerously after the rider has lost the reins. More and more, I
just do what I want when I want to do it. I don't notice anything awry
—and if I do, I don't care.

One hot and humid day in mid-June, I head to work in the early morning to avoid driving in heavy traffic because driving has become increasingly confusing. By late afternoon, I'm exhausted. I've worked
all day without a break, trying to make up for lost hours spent at doctors' appointments and hooked up to the IV to receive my immunotherapy drugs.

I look outside and see heavy, dark clouds gathering over the high-
rises of the NIMH campus. It's going to pour soon. I'm irritated by the
weather and so, so tired.

I have to leave. I have to leave right now.

I bolt from my office to the multilevel garage where I always park
and head to the same spot I always use. "My" space is usually open
when I get to work because I always arrive early, often before the garage has any cars at all. The garage I use is not the one closest to the
building where I work, but I like to take a little walk at the start and
then again at the end of my day.

For many years, I had limited need for these ugly, concrete con-
structions to park my car. Whenever the weather allowed, I biked
to work, about twenty miles each way on a peaceful, tree-lined trail
along the Potomac River. But no more. Since my brain surgery and
the immunotherapy, I don't have the same energy and stamina, so I
drive to work, although I hate it. I feel reduced to a lesser version of

myself. But at least I have a lovely walk to relax and unwind after a day in the office.

After ten minutes, I reach the garage. But I don't see my silver Toyota RAV4 in my regular spot.

That's weird. I don't remember having to park somewhere new today. I was in early, like I always am—wasn't I?

I walk up one aisle and down another. The garage is full, but my Toyota is nowhere to be found. I search each floor, trudging back and forth, scanning the rows of cars. I'm concerned, then very worried.

Someone stole my car!

Or maybe I just—I don't know. Maybe I parked somewhere new and don't recall it?

I reach into my purse and pull out the car key. I press the alarm button and hear a *beep*. It's coming from far off. I walk toward the sound, pressing the button from time to time to make another beep, then another.

What is happening? It makes no sense at all.

I retrace my steps, go back to where I started, and press the button on my key again. I hear the beep once more. But when I walk toward the sound, I can't hear it anymore. I try the same thing over and over: press, beep, nothing. I can't locate my car.

I'm confused and lost. I don't understand what is going on. I don't understand the world. It's playing tricks on me, strange and cruel tricks.

I see a woman walking in my direction. I hesitate for a moment before approaching her. How embarrassing to admit that I'm having trouble finding my car! But I have no other option. I'm tired of walking around in this dark space. I want to go home.

"Can you help me find my car?" I ask. "I don't know where I parked."

She looks surprised but says she will help. She takes my key, presses the button, and we hear the beep. "It must be halfway up on the higher floor," she says. "Look up there, through the gap between the floors."

There, in the opening she's pointing to, I see my silver Toyota. It seems to be on the ramp between the first and second floors. I have no idea how it got there. I grab my keys from her and run up the ramp to my car. It's flashing its lights, as if winking its eyes at me to say, *Gotcha!*

I'm relieved but confused.

Why is it parked here? I don't remember pulling into this spot. Is it possible someone moved it? Why would they do that?

My confusion only grows when I climb into the Toyota. I've been driving this car for three years, but when I get into it and try to fasten my seat belt, I can't seem to find it. I extend my hand as I always do to pull down the strap but there's no seat belt where I expect it to be. Instead, my outstretched hand dangles outside the door, hitting nothing but air.

I try again. The same thing happens. There is nothing to grab, nothing to hold on to. No seat belt, no anything.

Why am I having so much trouble with everything I try to do?

The world around me feels odd and awkward, with cars the most deceitful part of all. I no longer understand how to do the simplest things connected to them. I look around and still can't find the seat belt. Instead, I notice that my door is wide open.

It shouldn't be open, I realize. But I can't recall what that has to do with the missing seat belt. I sit for a while and then, irritated, I slam the door with a bang.

With that noise, my world returns to normal. Like magic. I slide my right hand across the inside of the closed door, and it easily locates the seat belt. I reach for it; it's in its normal position, right there, hanging from the fastener on the inside of the car. I pull it toward me and across my chest, slide the buckle into its locking mechanism, and click.

Finally! It works. I'm ready to go.

I start the engine and try to back out. But I get stuck. Something is holding the car in place. I can't move. I push harder on the gas pedal and hear a ghastly screeching sound of metal scraping something hard. I hit the brakes and look to my left. Somehow, I'm partially wedged under a small truck parked next to me. My wheel or some part of my car seems to be stuck under the truck but I am not sure how or why it got there.

I try to drive forward—the screeching intensifies. I put my car in reverse—the same thing happens. In desperation, I press really hard on the gas pedal, ignoring the terrifying noise of smashing and screeching and breaking objects, and finally free myself from the trap. As I pull away, I see that my car's left side is dented. But I don't check the damage to the truck. I don't care. I simply drive away.

I head toward the exit. It's clearly visible from a distance so I drive in that direction. Although the exit driveway is narrow and slightly curved, it's never given me any trouble. I've passed easily through it hundreds and hundreds of times. When I reach it today, though, it seems much narrower, almost unrecognizable. I drive slowly, trying to squeeze through the constricted exit. But I cannot fit.

What are they doing with these driveways? Changing everything, constant construction on this stupid campus! Why did they alter the exit?

I hear loud scratching and a bang as I run over a high curb.

The parking attendant runs out of his booth. "Lady, what are you doing?" he shouts.

"What do you think?" I mutter, increasingly irritated. "I'm just trying to get out of here, to leave this ridiculous garage and go home!"

Standing in front of my car, he points with his hands, directing my moves so I can free my wheels, as one of them is stuck high up on the curb. Finally, I pull loose. I drive off angrily.

I have the uneasy feeling that the world is plotting against me. As if to confirm that, as I head home, the sky opens and it begins to pour.

At this time of year in northern Virginia, the rains are often intense, almost tropical in their suddenness. Visibility during such weather is minimal; the world hides behind a curtain of water that is gray, foggy, and shapeless. Although the sun won't set for another several hours, it's dark and I see nothing but rain. I can't even make out the outline of the hood of my car. The houses, the highway railing, even the other cars, all seem to wash off in the rain. I'm driving blind.

Home is somewhere out there, a hidden oasis in the woods facing a quiet street. It's my cradle of safety. I need to get there quickly. Then I'll be fine. But it's nearly twenty miles away. I turn onto a busy four-lane road. Cars whiz by me at unusually high speeds.

Where are they going so dangerously fast?

I creep along to the correct exit and merge onto the main highway, the Beltway that snakes through the suburbs of Maryland and Virginia. From there it should be simple. I've taken this route countless times. But today it looks different.

Why can't I figure out where I am? Is it the rain that makes it so difficult?

I need the exit onto Little River Turnpike West. But I don't see it.

Have I already taken the exit? Why can't I remember?

Am I lost? I'm not sure. I don't really have any idea where I am. But I do see that I'm no longer on the highway. I keep driving. Instead of the familiar streets and houses of my neighborhood, I'm going past a huge shopping mall. Gray buildings, expansive parking lots, entrances to dark garages.

What am I doing here? How did I get into this gloomy shopping mall, someplace I've never seen before?

I feel as if I've skipped through time or leaped into another reality. It's odd. But I'm not worried much, and I'm not afraid. It's like I'm a character in a movie mysteriously transported in a rainstorm to a place I didn't intend to go. Nothing is what it seems. Nothing works as it should.

I want to get home but don't know what to do. I stop by the side of the road, then pull into a vast parking lot. I fumble with my cell phone. I know that I have an app that will guide me home but I cannot recall which one it is. I stare at the many icons on the screen but none of them are familiar. I randomly press the button on this one and that, but nothing's helping. After a long while I see the Waze icon and press it, and when it speaks its directions, I again begin to drive.

Eventually I pass by a large construction site with a building that extends along an entire block. It looks shiny and new and seems like it's almost complete. A huge sign announces that a Giant supermarket is soon to open.

A Giant! How wonderful! I wish they would build a new Giant near us!

Oh! Wait, look—it is in our neighborhood! I'm back in our neighborhood! This Giant will be ours!

My happiness quickly deflates. Yes, this will be our new neighborhood grocery store. But will it be *mine?* Will I live to see it open?

Now I'm in my driveway. I have no idea how I got here.

It's becoming more and more difficult for my brain to function normally. Increasingly, I find it a struggle to carry out ordered, sequential movements. I can no longer execute simple tasks that I've done many times before or organize them in my mind in a methodical fashion. By itself, each step is very familiar, but combined together, they are as challenging as the complex experiments I used to carry out in my labs. I know that I can't drive without wearing a seat belt, and I vaguely know where the seat belt is supposed to be. But I can't perform the basic steps to buckle it—steps that were automatic for me only days ago.

What part of my brain isn't working? It's likely that communication between my prefrontal cortex and my hippocampus is failing, which is unpleasantly reminiscent of the prefrontal cortical connections I disrupted in rats to study schizophrenia. Perhaps the areas that aren't functioning normally could be determined if I were to undergo a battery of neuropsychological tests as my problems intensify. But no one is testing me like I tested my rats, in controlled experiments carefully designed to examine particular elements of the behavioral impairments. Still, I share some similarities to my brain-damaged rodents: I can't find my way in the maze of streets in my cozy neighborhood, and I cannot locate the sweet rewards of food and safety that are waiting for me at my destination.

In some ways, my struggles are similar to those of people who have a condition called dyspraxia, the loss of motor skills, motor memory, and the ability to perform coordinated movements. Dyspraxia can be due to a developmental disorder; the actor Daniel Radcliffe has been open about his struggles with it. Dyspraxia is also quite common among people with Alzheimer's disease, and the symptoms can be progressive: First, patients have trouble with sophisticated motor skills, and later, they are unable to do simple things such as brush their teeth. Eventually, some can't even swallow.

These kinds of difficulties are also common in people with damage to the parietal cortex. The parietal lobe is also related to the ability to read and do math; dyspraxia often coexists with dyslexia and with dyscalculia (difficulty doing arithmetic, something I will soon begin to experience as well). This could have suggested, if we'd thought about it at the time, that my brain problems were more widespread than any of us imagined.

In addition to dyspraxia, I'm also suffering from loss of visuospatial memory, which makes it hard for me to remember my location and navigate my way through space. These problems are similar to those described in people with a condition called developmental topographical disorientation (DTD). From very early on in their lives, perhaps from birth, people with DTD don't recognize very familiar environments. Just as I could not find my way home in a place where I've lived for almost thirty years, people with DTD have no recognition of their surroundings no matter how many times they follow the same route. For me, this was short and transient; for them, it is permanent.

Spatial orientation involves multiple regions of the brain and a network of connections among neurons in different areas. Two regions, however, stand out as crucial for spatial memory: the prefrontal cortex and the hippocampus. In the case of DTD, it may be the connectivity between these two regions that has gone awry, as MRI scans have demonstrated to neuroscientists who study this rare neurological disorder.

Is this what is happening with me? It's possible. My prefrontal cortex appears to be dysfunctional and perhaps can't connect efficiently with other brain regions, including the hippocampus, one of its main, albeit indirect, targets. It's possible that the lack of communication between these two regions in my brain is the reason why I cannot figure out where I am, even when driving through a neighborhood where I've lived for decades.

My changed behaviors do not raise enough red flags for my family and colleagues to wonder whether my brain may be seriously malfunctioning. For one thing, I'm not telling my family the whole truth about every single problem I encounter. I don't even tell them about how I smashed the car. The lapses in how I usually act can be, and are, easily discounted and explained by the stresses of my grim diagnosis, challenging treatment, responsibilities of family and career.

And, regardless, I'm still functioning at a very high level—which is remarkable, given what my family, my doctors, and I are about to learn about the shocking reality of what's happening in my brain.

7

Inferno

*T*his headache is killing me.

Dull and throbbing like distant thunder, it overwhelms me, taking over not only my head but my entire being. The clock in my bedroom shows that it's the middle of the night. I lie in bed wide awake.

Somewhere deep inside my body I feel an approaching storm. All of a sudden, lightning strikes. My stomach turns upside down, nausea overcomes me, and I leap from bed, rush to the bathroom, lower my head over the toilet bowl, and vomit violently. My skull seems to split in half as the headache explodes, then slowly recedes. I feel better but so weak I can't stand up. Kneeling before the toilet, I stare down at strange pieces of plastic swirling in the water.

I am terrified. It's surreal, the sight of all this plastic I've vomited up.

Why would they have made a pizza filled with plastic? Poison. They are poisoning us!

Last night, June 16, we celebrated my final infusion, a finish line I'd sworn I would cross. I was elated but very tired. I felt as if I'd just graduated from college with the highest grades in my class or crossed the finish line of a marathon. Done with immunotherapy! After twelve weeks of hoping that I could withstand the hardships of this treatment—itchy rashes covering my body, gastrointestinal problems, the loss of thyroid function—it was over. That final hospital visit was the longest ever—over six hours waiting for blood tests, waiting for the doctor, and waiting for the drugs to be delivered from the pharmacy in their transparent plastic bags and then released very slowly, drip by drip, into my veins. Afterward, Mirek and I were both so tired that I couldn't even consider making dinner. On the way home from the hospital, we did something we rarely do: we stopped for takeout pizza at a local restaurant.

We don't go to restaurants or get takeout very often. We both prefer my cooking, one of my great pleasures. In America, I've taken advantage of the freedom of a previously unimaginable selection of foods by cooking as often as I can. I've prepared our meals for years, no matter what my day was like, no matter if I was going through chemo for breast cancer or recovering from a mastectomy or brain surgery. After every marathon and triathlon that I've raced, I've returned home tired but beaming with happiness and prepared our dinner. Usually it's something simple and healthy: pasta with stir-fried vegetables and grated Parmesan; baked fish, roasted potatoes, and arugula salad; chicken stir-fried with sweet peas, tomatoes, and onions, and spiced with lots of hot red pepper. Mirek and I love to sit in our spacious dining room overlooking the woods enjoying glasses of wine—or, more often, a bottle. We share the events of our day, relive a road race, discuss what I talked about with Kasia or Witek or Maria in our daily conversations. This is our sacred time for relaxing and catching up with each other. Our dinners last at least two hours. To mark the end of the meal, we drink strong, hot tea.

Now, as I stare at the bits of plastic in the toilet, I regret the decision to break from our ritual.

The restaurant filled the pizza with plastic! Pieces of plastic bags! To make the pizza look bigger so they could charge more money! I should have known! The cheese was so white, weirdly white, with a crumpled texture too strange to be real food. It didn't taste like real, crunchy pizza. The bottom was soaked in some kind of odd liquid. And the top! Covered with chewy, inedible plastic!

I am seething. We've been poisoned!

"Mirek! Wake up!" I rush into the bedroom. "The pizza! It's poison! It was made of plastic!"

He sits up in bed and tries to calm me.

"It isn't poison," he says gently. "The pizza wasn't that good, but there was no plastic or anything like that in it."

"No, listen to me," I say. "I just threw it all up. The pizza was made of plastic! I saw it floating in the toilet. The cheese was plastic, the crust was plastic."

"But I didn't get sick," he says soothingly. "Don't you think your vomiting was a reaction to yesterday's infusion?"

"Don't you believe me?" I grow more agitated. "I saw it. I saw the plastic. They are poisoning us!"

He gently pats my back, asks if he can get me some water. "Come lie down, try to sleep," he urges. "You'll feel better."

I announce we will never eat there again. Mirek agrees. But as he falls back asleep, I lie beside him, angry and suspicious.

Why won't Mirek see what's going on? Why is he defending the pizza place?

In the morning, I call Kasia and tell her that the pizza place down the street tried to poison us with plastic.

"Mom," she says carefully, "I think you should call Dr. Atkins or his nurses." I can hear the concern in her voice. "Please call them."

"It's not me! It's the pizza place!" *Why won't Kasia believe me?*

"Mom? Will you please call them?" she presses.

"No, no, I'm fine," I say. "It was only that awful pizza. Never mind. It's already passed."

On Wednesday and Thursday, I drive myself to the office in the morning and spend uneventful days at the brain bank. Thursday after work, I go swimming at the local pool, then go food shopping. When I get home with the groceries, I tell Mirek that I'm feeling very well. But after dinner, when I sit at the computer to continue writing my life stories, Mirek notices that I'm having trouble typing. He also sees that I have no idea how much I'm struggling; I don't notice that some of my words are mangled. Mirek says nothing to me but heads upstairs to telephone Kasia. They talk about the incident with the pizza and the splitting headache I had that night. They're very concerned about my behavior.

Early the next morning, Friday, Kasia calls me.

"I really think you should contact Dr. Atkins," Kasia says. "I'll draft an e-mail to him and send it to you. You can forward it to his nurses."

A few minutes later I receive the note that Kasia wants me to send:

My daughter wanted to bring this up, though I feel fine. She is worried that there may be subtle changes in my driving and perhaps in thinking (mild forgetfulness, forgetting to turn at the right intersection). This could be from stress, feeling down, or something else. Given the ongoing headaches and especially the severe headache I had the other day she is concerned about swelling or inflammation around the brain lesions. Can you bring this up with Dr. A. and see what he thinks. Many thanks.

I am furious. My own daughter is betraying me.

Kasia is a very smart physician, and I know that she's upset and worried about me. But she is being hysterical and irrational. And she's really overstepping her bounds! As if there's something wrong with *me!*

I have my own very good mind, and I have much more life experience than she does. Everyone in the family respects my intuition and
judgment, not only about my own well-being but about the health of
all of us. Kasia may be an experienced doctor, but she calls *me* when
she doesn't feel well. She calls *me* when her kids are sick, and not just
to share her worries and seek comfort. She always wants my advice.
*Mom, do you think it's serious? Should I call the pediatrician? What
if the fever gets worse? What if…* I always tell her what *I* would do,
and more often than not, she follows my advice. I am still her wise,
trusted mother, after all. So why is she treating me like this?

I e-mail Kasia back:

> I will not write it, maybe I will call the doctor, but please do not tell
> me what to do. Mama is in charge of her own fate and will do what
> she thinks is appropriate. I know you are worried, it is moving, but
> please, leave my decisions to myself. I am fine!!!

Moments later, Kasia responds with an e-mail:

> Mama!!!!! Ok!!!! I respect your decisions and will do as you wish.

I don't call the doctor. A little while later, Kasia calls me and offers again to contact them herself. For whatever reason, I don't object anymore. An hour later I get a phone call from Dr. Atkins's nurse,
who says she received Kasia's e-mail and wants me to come immediately to the hospital. She's scheduled an emergency MRI for an hour
from now.

"Let's go and get the MRI," Mirek says. Even though he isn't push-
ing me, something about the way he speaks makes me suspicious.

*Why is Kasia plotting against me? Mirek is on her side too! They're
all against me!*

I'm still annoyed but I agree to go. I pick up my car keys and head
outside.

"But you've been having some trouble with directions. Why don't
you just relax and let me drive?" Mirek suggests.

"I always drive!" I retort, and I climb into the driver's seat. He reluctantly gives in.

No sooner do we get onto the highway, however, when he starts shouting: "Watch out! Look out!"

What is he going on about?

"You're not inside the lane!" he cries. "Stay in the center! No, no, you're crossing over the line again! Pull back, pull back!"

"I'm fine!" I insist. "It just looks different from where you're sitting. Why are you so critical of me all the time? Can't you just be quiet?"

But cars behind us start honking, and I realize I'm about to hit the truck to my left. I swerve sharply at the last minute. Mirek has his head in his hands.

"Oh, stop it," I say. "Nothing happened. It's not a big deal. Get over it."

With no further drama, we check in at the Georgetown MRI center. A nurse inserts the needle into a vein in my arm so that contrast liquid can be injected. I lie on a narrow table, and a technician slides me into a powerful magnet that looks like a tight tube. With my head secured in a plastic crate and my body wrapped in white blankets, I look like a mummy.

I remain motionless while the magnetic field is switched on and off to the accompaniment of loud tapping noises from vibrating coils that are invisible to me. Unable to see anything in this tunnel, I am alone with the mangled thoughts in my confused brain. The *knock-knock-knock, knock-knock* sound of the MRI machine, repeated over and over in varying rhythms and pitches, is strangely relaxing. I like the solitude. I feel cozy and safe, happily cocooned in this tight space. It shields me from the nonstop stimuli of the outside world.

After an hour, the MRI wraps up. I dress and find Mirek waiting in the hallway.

"Done," I say. "Let's go home."

Before we reach the parking garage, Mirek's cell phone rings.

"What? Why?" he says. "Oh, okay, we'll be there right away."

He turns to me and says, "We have to go immediately to the emergency room."

"Why? What's happening?"

"The nurse said your brain is very swollen," Mirek says.

As we walk over, I realize my headache has returned, insistent and intense.

In the ER, they take me quickly to a back room and check my blood pressure. It is very high. They lead me to a cubicle, where I lie down in bed behind a curtain amid the awful noises of trauma and emergency. People outside my cubicle run and shout and cry and scream. Here I am again, just five months after they discovered the bleeding tumor in my brain.

But I'm not worried in the least. In fact, I don't fully understand why we're here. Mirek's eyes are sad, his face troubled, but I can't fathom why he's upset. I try to cheer him up, try to joke. But his expression doesn't change. He just holds my hand and looks at me.

After a while, my oncologist Dr. Atkins enters my cubicle with two of his nurses. They look at me with such sorrow that I think some kind of a mistake has been made. They can't be worrying about me —why should they?

"The MRI shows new tumors in your brain," Dr. Atkins says. "The immunotherapy didn't work. I'm really sorry."

I look from face to face. Mirek is somber. Dr. Atkins seems deeply disappointed, as if he's failed me.

My poor doctor. He doesn't understand—I'm fine!

"There's also swelling and serious inflammation of brain tissue," Dr. Atkins continues. "I'm prescribing high doses of steroids right away to reduce the swelling, and I'm admitting you to the hospital."

Oh, Dr. Atkins—I feel so sorry for him. Let me reassure him.

"No, no, please, wait," I say. "I don't want steroids. From what I've read, steroids will reduce my immune response and interfere with my treatment. And I *know* the immunotherapy worked. I *know* it. I'm

sorry about this inflammation in my brain but you know it can happen. There are often setbacks with immunotherapy before there's
improvement. Don't worry, please. I will be fine."

I look at Dr. Atkins, then at Mirek, whose eyes are filled with tears.
The two nurses also look as if they are about to cry.

*All of this fuss for no reason! Let me explain to them what's happen-
ing—maybe that will calm them down.*

"Tumors often get larger at first when this treatment begins," I
say. "I remember that, I swear I do, from several scientific publications I read just weeks ago. The tumors you see on the MRI may look
larger than they really are because my T cells are fighting the melanoma cells and killing them. What you're seeing is evidence of this
dramatic war in my brain. We have to give my body time to clean up
this ugly battlefield. We just have to wait. Trust me."

But Dr. Atkins shakes his head. They all look through me, past
me, their eyes glistening, their faces solemn. They talk among themselves, not really listening to me. They lean over my bed, examine my
face, and worry.

I feel so sorry for them. I wish they could understand that I'm
right.

Mirek tells me Kasia is on her way from New Haven, and she arrives a few hours later, joining us in the hospital room to which I've
been moved. I am stunned to see her. "Kasia, oh, baby, you shouldn't
have done this! I'm really okay," I assure her. She begins to cry. She
has canceled her Italian vacation with Jake and the boys—something
they'd planned for a year—to rush here. I'm happy she's with us but
flabbergasted at her decision and her outpouring of emotion.

"This is a lot of excitement for no reason," I tell her. "I am fine! I
am fine!"

By now it's almost night, and Kasia climbs onto the bed with me
just as she did in January, as tired and upset now as she was then.
It feels good to have her so close but I still don't understand the ur-

gency. I don't know how to persuade her and Mirek and Dr. Atkins that there's really no reason to fret so much.

A few hours later, Mirek and Kasia go home, telling me they'll return in the morning. "Of course!" I say cheerily. "I'll be fine. I don't really need anything. Don't worry and don't hurry—have a nice bike ride in the morning." I don't have a toothbrush with me or even a change of clothes, but I'm feeling upbeat and well. The headache is gone. Hours later, I send them a selfie that shows me smiling in my hospital garb in bed.

The selfie I sent to my husband and daughter from the Georgetown hospital.

But I don't have a restful night. Nights in hospitals are never good —so much commotion and noise, bright lights and beeping machines. I'm awakened at dawn by a nurse checking my vitals and changing the bags on the IV pole. I'm angry to be roused from sleep, and hungry—so, so hungry.

"When is breakfast?" I ask.

"Soon," she replies.

"But I'm hungry!" I respond. *I'm hungry. I want to eat.* It's my only thought.

At seven o'clock, my breakfast still has not arrived. Nor is it there at eight or nine. I'm fuming. When the nurse reappears, I pounce.

"How is it possible that breakfast hasn't been served yet?" I scold. "What a terrible hospital this is. My insurance is paying hundreds of dollars a day for me to be here. Appalling! Breakfast itself will probably cost a hundred dollars—and it's late!"

I repeat my complaint to everyone who comes into my room. Ten o'clock arrives, and still no breakfast. And no Kasia or Mirek either. When they finally telephone, I let them know I'm furious that they're not here yet with some food. After we hang up, I walk into the nurses' station, dragging my IV pole, and demand my meal. The nurse explains that because I'm a new patient, it takes longer than usual to order my breakfast. I storm off, stop a doctor in the hallway, and insist he hear my tirade: "No breakfast! How pitiful, how irresponsible. My insurance is paying for it!"

No one can escape me, not the nurses, not other patients. They all need to hear my breakfast story, and I make sure they do.

Finally, at ten thirty a.m., the hospital staff brings my breakfast, just as Mirek and Kasia arrive carrying oatmeal with fruit and nuts, my favorite morning meal. First I devour the hospital food, then the treats from my family. But I remain unhappy. I repeat to Kasia and Mirek the story of the tardy breakfast, and retell it, and tell it again. Every nurse and doctor who comes into the room is greeted with the story of this injustice. They try to ask me about my headache and other medical issues. But I want to tell them that my food was late. And I'm still hungry! Can't they bring me more?

I see that my daughter is cross. She tells me to stop. "Mom, don't you understand that you are seriously ill?" she says, her eyes welling up. "You have new tumors in your brain. Why are you stuck on unimportant things like breakfast and food when your life is in danger?"

I can't believe what I'm hearing. "Breakfast, unimportant?" I retort. "Of course it's important! It's important for *me.*"

Kasia leaves the room. I can hear her just outside the door, talking to a doctor who has come to examine me. When she reenters, she's crying. Her emotional reaction puzzles me.

"Why do you want to talk about tumors and sad things like that?" I tell her. "What's the point? What can I do about it? You're overreacting."

"Mom, you're very sick," she responds. "Don't you get it?"

"You're panicking. Calm down!" Then I add, "The whole world is against me!"

"I don't recognize you! You're not the mother I've known all my life!" She continues sobbing.

I stare into the distance silently.

Nobody loves me anymore. I simply can't believe they don't agree with me about this breakfast debacle. Breakfast at ten thirty? For what we're paying?

At the hospital, I continue to polish off everything on my meal trays while asking my family to bring more food from home. I find the hospital crackers especially alluring. I gobble them up and look for more. Everything is delicious.

Midafternoon the next day, Sunday, June 21, they release me from the hospital. I'm going to continue on high doses of oral steroids, and I have an appointment with Dr. Atkins in a few days to hear details about my condition and discuss options. Until then, we're just waiting. Nobody in the family mentions the possibility of further treatment. Death hovers among us like a ghost.

When we get home, I'm still very hungry, and I insist on making dinner. But I'm baffled as I try to prepare the meal. I can't find the pots and pans or anything I need. When Mirek offers to take over, I tell him to leave me alone. Kasia also tries to help me but I criticize her efforts so much that she, too, retreats. We three eat dinner in near silence.

Over the next few days, I find it harder and harder to prepare our meals. I have no idea how to adjust the measurements in recipes I use for just Mirek and me in order to make enough for Kasia, too. I also forget the proportions of ingredients in even the simplest recipes: how much water in relation to pasta, how much salt to add for that much water. And I completely lose the ability to plan; I can't figure out which dishes need to be cooked first in order to time the meal appropriately or even which ingredients to add to which recipe and when. I can't even bake bread anymore, a ritual that I've been performing every week for years, using a starter yeast from Poland. No matter how much I try, I can't remember how to do it.

Yet as frustrated as I am in the moment, I do not reflect on what it all might mean. It's as if I don't remember that just weeks before, I was good at these things. I make no connection between the serious problems in my brain and the fact that I can't cook my favorite dishes.

Even as I struggle to function in the kitchen, my obsession with food continues. I will gain ten pounds between the middle of June and the beginning of July. And it won't bother me at all. When I emerged from brain surgery in January, I was very thin; I weighed 118 pounds, the thinnest I'd been in adulthood. But soon I'll soar to 138 pounds, more than I would normally think of carrying on my five-foot-six-inch frame. And I don't care. Steroids often cause people to gain weight, but that's only part of the problem. I simply cannot stop myself from eating. It's not about being hungry, it's just that these treats look so good that I'm going to eat them! Why not?

Worried about the health consequences of so much sugar, Kasia gently suggests that I might want to try taming my ravenous appetite.

An endocrinologist, she's particularly concerned because I'm on ste-roids—steroids combined with too much sugar can cause hypergly-cemia.

"Mom, please," Kasia says. "You don't want to be eating *all* of the ice cream, do you?"

"Leave me alone," I retort. "You can't tell me what to eat. It's my business, not yours."

None of us see it at the time, but my obsession with eating is a clas-sic sign of frontal-lobe problems, which in my case are compounded by steroids that by themselves increase appetite. People with fron-totemporal dementia often gain significant amounts of weight very quickly, since they have no inhibitor on their drive to eat. When the frontal cortex is operating as it should, people have the ability to weigh the pros and cons of fulfilling their desires. But when that function is silenced or gone, they just do what they want with zero concern about the consequences.

I love sweets so I'm going to eat them—period!

On Wednesday, June 24, Kasia, Mirek, and I return to Dr. Atkins's of-fice to learn what's next for me. I'm curious to hear what he has to say. The steroids are giving me so much energy and I feel so much better that I know I'm on the mend, new tumors or not.

I smile at the receptionist as we check in. But Kasia and Mirek aren't in good moods. They sit solemnly in the waiting room until Dr. Atkins's assistant comes to get us.

"Hello!" I say to her brightly. "Very nice to see you again!"

She gives me a sad, fleeting smile as she leads us to the examina-tion room.

When Dr. Atkins enters, his face is grave. He asks us to take a seat. Three of his nurses, Kellie, Bridget, and Dorothy, stand near him, looking heartbroken.

"Good afternoon!" I say cheerfully, trying to lighten the mood. "How bad can this news be?"

"As you know," Dr. Atkins says, "there are new tumors in your brain—"

"We'll simply have to deal with this," I interject. "I've had new tumors before. They'll eventually shrivel and go away, believe me."

Bridget, the nurse closest to the door, is failing in her efforts not to cry. She turns her face from us and wipes tears from her cheeks.

"Really! It's fine!" I assure them. "I'm telling you—"

"There seem to be at least eighteen tumors in your brain," Dr. Atkins says.

Kasia gasps.

"As you know, you had three brain tumors when you entered the trial," Dr. Atkins says. "About fifteen new tumors have emerged throughout your brain since the last MRI."

"*Eighteen?*" Kasia says, her voice breaking. Mirek tenses beside me but says nothing.

"Oh, I don't think that's the case," I say. "What you're seeing is something else, inflammation perhaps, or some—"

Dr. Atkins interrupts and offers to show us my scans in his office next door. Kasia goes out with him but I don't follow them, and Mirek stays with me. When they return, Kasia's eyes are glistening.

The scans show a scattering of small but distinct black spots in my brain, Dr. Atkins tells us—over eighteen little shapes the size of raisins. The largest tumors are in my frontal and parietal lobes, he says, but they also lurk in the temporal and occipital lobes, and in the basal ganglia, a group of brain structures in the base of the brain that help coordinate movement. Later, Kasia will tell me that on the scans, my brain looked like a lump of raisin-bread dough.

The largest tumor, Dr. Atkins says, is in my frontal lobe. It's the size of an almond.

"No wonder you have been acting so unlike yourself," Kasia says quietly.

"Really, Kasia—I haven't been acting that different!" I say.

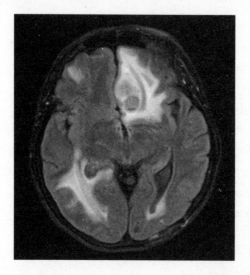

The brain scan, done on June 19, in which Dr. Atkins discovered new tumors and extensive swelling. The white areas show swelling; the tumors are circular blobs. The clearest of these, in the upper half of the image, sits squarely in my frontal cortex.

Dr. Atkins nods to Kasia and continues. "And the scans show a number of fuzzy, whitish areas, which means that a large part of your brain is very swollen."

"Mum, I love you," Kasia says in Polish.

"But these steroids will stop the swelling! I'm already feeling better!" I say, my smile broadening.

I look to Mirek, who is staring at me. I look at the nurses, all of whom are tearing up again.

Why are they all so pessimistic? They're overreacting. This doom-and-gloom is so unnecessary.

"I'm sorry the immunotherapy didn't work," Dr. Atkins says once more. "I was so hopeful that it would."

No one else speaks. A heaviness pervades the room. But I will not give up.

"Well, okay, what's next?" I ask. "What do we do?"

"We'll do radiation therapy of the tumors," he says. "Our radiation oncologist, Dr. Sean Collins, will contact you shortly."

But we all know radiation is no cure.

"And then?" I ask. "What if that doesn't work?"

Dr. Atkins hesitates.

"Please, just tell me," I say. "What comes next for me?"

I feel detached from my question, like a scientist asking about a specimen in a jar, as if what we're discussing has absolutely nothing to do with my own mortality.

"As the swelling increases and places more pressure on your brain, you will likely fall into a coma," Dr. Atkins says.

A coma? A coma doesn't scare me. It sounds comforting, like sleep.

"And then?" I ask.

"And then—you will eventually die," he says quietly.

"Okay," I say. "In the meantime, what should I do? How should I prepare?" I ask this as matter-of-factly as if I'm seeking advice on weatherproofing the patio.

He seems unsure how to respond. Finally, he says, "It's time to get ready for the worst. You should get your affairs in order."

Everyone else in the room is choking back tears.

I don't feel like crying at all.

"Okay." I nod. "I like a plan of action. I will get my affairs in order." Then I immediately realize that I don't need to, actually—I straightened out my affairs months ago, when I got the brain cancer diagnosis. The fact that I'm fully prepared gives me a renewed sense of calm and satisfaction.

Everyone else looks devastated.

They're all so upset. But I'm fine. They'll see, I'm fine.

We don't speak another word about death. On the car ride home, Kasia, Mirek, and I don't talk much at all.

I sit in the passenger seat, going over in my mind what I've read in the scientific literature about immunotherapy. I'm convinced that this swelling in my brain, these new tumors, are a temporary stage in what will end up being a successful treatment. I recall what the research describes about some cases—the tumors swell, then they shrink and disappear. My ability to remember what I've read about my treatment isn't failing at all, and it's keeping me optimistic.

From my long experience studying schizophrenia, I know that brain problems lead to poor judgment and an inability to recognize one's own mental deficits. But at this moment, all my years of professional expertise aren't helping me see things as they really are: I'm losing my mind—and my life.

Several days later, on Sunday, June 28, Kasia and I stand in the local Home Depot.

Blue. Orange. Pink. Red. White.

Impatiens of every color are arrayed before us under the awning in the garden section.

"Mom, we've been here fifteen minutes," Kasia says. "Just pick some."

I cannot make up my mind. How many do we need? What colors do I want? I like coral but none of the choices are close to that hue as far as I can tell. Is this coral? I'm not sure. Maybe. But these plants don't look fresh. Kind of wilted. Okay, then maybe not coral after all. Maybe red.

Kasia sighs in frustration.

I can't decide. I give up. After half an hour of closely inspecting flowers, I settle on something purplish, or reddish—I'm not sure. We get in my car and Kasia drives us to an Asian restaurant in a nearby strip mall for a special treat: sushi takeout for Mirek's birthday.

Forty-five minutes after we left Home Depot, I'm sitting alone at the counter in the restaurant. All around me, people rush about speaking loudly in languages I don't understand. It's lunch hour, a busy time for this casual place. It's filled with people from all over the world, particularly Korea, the country of origin of the most recent wave of immigrants to these northern Virginia suburbs. For some reason, right now I find the commotion entertaining.

It's a pleasant distraction, because I'm stuck. I'm trying to think about something, but I'm really struggling. It's hot outside, and hot and stuffy inside the restaurant. The air is filled with exotic aromas—kimchi and steaming plates of noodle soups, marinated meats such as bulgogi grilled right on the tables around me, garlic, ginger, and soy sauces. Such a far cry from our bland Polish cuisine of pierogis and cabbage and meats stewed for hours with onions and wild mushrooms until they resemble a brownish, mushy pulp. Our family mostly did away with those foods years ago. In honor of our Polish tradition, we eat them during the holidays, when we savor them for the nostalgia they bring.

Mirek chose sushi for his birthday dinner; it's his favorite. I'd almost forgotten that tomorrow, June 29, is his special day. When I called my eighty-seven-year-old mother in Poland this morning for our weekly talk, she'd asked, "Is Mirek's birthday tomorrow?" I couldn't remember. I knew this time of year was important in our family because we celebrate two birthdays, Mirek's and Ryszard's, my brother-in-law. But whose birthday was coming up? I didn't know. "I think so," I'd answered vaguely.

To be sure, I had to check with Kasia. "Is tomorrow Ryszard's birthday? Or Mirek's? I can't remember."

"Tomorrow is Mirek's birthday," she'd said. "Ryszard's was a few days ago."

I should have been surprised that I couldn't remember the birthday of the man to whom I've been married for almost thirty years, a man I love with all my heart. For years, his birthday has also been the code I've used to unlock my cell phone. But I'm not easily surprised nowadays. There are lots of things I can't remember. Numbers, in particular, escape me, and dates are difficult too.

Since I'm going in for radiation tomorrow, Kasia and I decided to celebrate Mirek's birthday a day early. And now, as I sit here in the restaurant, I keep staring straight ahead. The waitresses are curious as to what I'm doing, I can tell. With kind smiles, they ask if I need anything more, if they can help. I thank them but shake my head. The sushi chef across the counter from me, a tall, handsome man, works on a sushi roll, cutting and chopping colorful ingredients, rolling sticky rice wrapped in seaweed with his bare hands, squeezing fancy sauces on top. As he dips his fingers in this and that container, he glances at me with a shy smile.

It's been twenty minutes since they delivered our takeout order to me, a big brown bag with a tray of creamy sushi rolls filled with eel, salmon, and whitefish decorated with avocado, wasabi, seaweed, sesame seeds, and other spices. I'm still trapped here at the counter, staring at the bill and trying to figure out the tip.

I've made no progress whatsoever. I see lots of numbers scrawled on a small sheet of paper but they mean nothing to me. I read the numbers but don't understand what to do with them. I do remember that a tip should be 20 percent—that concept springs to mind—but I don't understand the notion of percentages. I remember only a bare fact: 20 percent. Without more context, it is meaningless. What does 20 percent signify? How does one calculate it?

I scrutinize the bill. What was the cost of our sushi? I think that's it, that number there, seventy. But if that's how much the food cost, how much is the tip?

I turn these questions over in my mind, desperate to find an answer that doesn't come. I change my strategy and start playing with random numbers in my head, then try them on my tongue. "Thirty dollars?" I whisper. "Or twenty dollars? No, that doesn't sound right."

I throw looks in the direction of the restaurant's front door, through which Kasia disappeared nearly half an hour ago. She went to get the car, I remember, so we wouldn't have to carry the tray too far.

Why doesn't she come back?

I feel helpless. I open my wallet and find a ten-dollar bill.

Okay, maybe ten dollars.

I resign myself to the amount that's available, a random bill that I place on the counter. I quickly leave so I won't be stopped and questioned if it's wrong. I feel like a crook.

Kasia has been sitting in the car near the entrance to the restaurant all this time.

"What happened, Mom? What were you doing in there for so long?" she asks.

I don't know how to reply. "Oh, nothing," I say, trying to sound nonchalant. "Do you think ten dollars is okay for the tip?"

"You left a tip on a takeout order?" She sounds surprised.

"Why not? But I had trouble calculating the right amount."

She gives me a bewildered look. "How much was the sushi?" she asks.

I hesitate. "Seventy dollars," I say. I feel relieved that I can recall.

"You couldn't calculate twenty percent of seventy dollars?"

"No." Suddenly, I feel inadequate.

As we drive home, she starts quizzing me. "What's a hundred and twenty divided by three?"

I ponder it. "I don't know," I say.

"What about twelve divided by three?"

"I—I have no idea."

"Can you add five and ten?" she tries.

"Fifteen!" I shout, overjoyed.

"Eighteen minus five?"

"I don't know. Twelve, maybe?"

We experiment with simple arithmetic problems all the way home. We discover that I can add, as long as the numbers are simple. But any kind of subtracting, multiplying, or dividing is impossible, no matter how basic the question. These calculations are simply over my head.

Celebrating Mirek's birthday dinner with his favorite food, sushi. I'd just discovered I couldn't calculate the tip for the sushi or do other kinds of simple math.

When we walk into the house, Kasia and I don't talk about it anymore, and we don't mention it to Mirek as we celebrate his birthday with our sushi meal. Kasia doesn't tell me until much later, but it deeply pained her to see me so deteriorated, so altered, from the strong-minded and accomplished person I used to be: her sharp-witted mother, the one who taught her math and logic as well as the importance of honesty and how to enjoy her life. She doesn't want our roles to change. She doesn't want to be a physician examining my symptoms and observing my strange new behaviors in an attempt to understand what's wrong. She wants her loving, fun, competent mama. Not this confused, angry, self-absorbed impostor.

• • •

As Dr. Aizer will explain to me much later, my compromised math ability—which is called dyscalculia or acalculia—is most likely related to the lesion and inflammation in my parietal lobe, the area located just behind the frontal lobe on top of the brain. Together, the frontal and parietal lobes make up about two-thirds of our species' highly evolved neocortex, which comprises the four lobes of the brain. Lesions or defects in the frontal and parietal lobes have been linked with dyscalculia in patients with early stages of dementia.

Scientists have been able to trace different aspects of numerical processing, such as multiplication and subtraction, to different subregions of the parietal lobe. Thus, people with lesions in a particular area of the parietal lobe may show deficits in the ability to perform one type of calculation but not others. In my case, I seem to be able to add simple numbers. But I cannot handle division, subtraction, or multiplication. It may well be that the swelling in my brain is affecting the function of specific subregions of the parietal lobe while leaving others relatively unscathed.

Lesions in my parietal lobe, which Dr. Atkins pointed out to us during our last visit, may be giving rise to other problems I'm experiencing. The parietal lobe has a role in processing topographic memory, the ability to recall the shape and structure of a previously experienced place or to hold in one's mind the map of a place. It's also involved in motor planning, the ability to plot and execute skilled tasks that aren't habitual. And it's involved in the ability to have insight into one's illness, which I clearly lack. For me, all of these functions are now impaired.

Incredibly, though, my writing ability is not deteriorating at all; if anything, it's getting stronger despite my short-term memory problems. Not only are my language skills still intact, they are also surprisingly robust. Perhaps fueled by the steroids, my creativity is on fire. I awaken at four or five o'clock each morning and prop myself up in bed, my laptop on my knees. My mind is swirling with ideas on how to describe what I feel. My emotions and memories are so intense, and sometimes so bizarre, that I have to turn them into words, both to unload the burden and to share these vivid recollections with others before they fade. It's as if I'm compensating for the shortfalls of real life by transferring them onto virtual paper: a computer screen.

I write about my childhood in Poland, about my beloved grandma taking us into the primitive, remote villages of the Beskid Mountains on our summer vacations. With an overpowering delight, I retrieve long-forgotten memories of the smell of sweet hay and cow manure. I gather mushrooms in the woods, cross icy streams, and pick wild blackberries with my grandma and my little sister. These memories from more than fifty years ago are so vivid and pleasurable that I don't want them to disappear. I type pages and pages recalling when my sister and I were little girls living in a world far away. I remember it as clearly as if it happened yesterday.

When Maria comes to visit in July, I share these memories with her. She's amazed and delighted at the incredible details I can recall from our earliest years. But I also detect that reminiscing about our childhood makes her sad, although I don't know why. Only later will I realize that she, like everyone in my family, is traumatized by the prospect that soon I will cease to exist, and memories of me will be all that will remain.

Throughout July, family members take turns visiting me: my sister and her husband, then Kasia, then my son and Cheyenne, then Kasia again. They keep me company, and I like it. It's flattering that they pay so much attention to me but they're all anxious and somber. I sense that something is horribly wrong and that's why they're rotating through here so often, but I can't figure out what they're worried about. Since I began taking high doses of steroids I don't have headaches anymore, which is a great relief. I am upbeat, unmoved by the latest news of the multiple new metastases in my brain.

Tumors. More tumors. Oh, well. What I should make for lunch today?

I'm feeling almost happy. And I would feel even better were it not for the uneasy sense that my family knows something I don't—some kind of a tragic secret that I don't fully comprehend.

8

Chanterelles

A *week after being released from* Georgetown University Hospital, I return as an outpatient to undergo radiation treatment to most of the fifteen or so new tumors in my brain, including the ones that weren't radiated before I started the clinical trial. For the time being, only the two smallest—tumors too tiny to target—will be left untreated.

This is the first time I will undergo the CyberKnife procedure. As opposed to the stereotactic radiosurgery that I received in March after neurosurgery at the Brigham, the CyberKnife Robotic Radiosurgery system is almost fully automated. Now, as I did in March, I lie strapped to a gurney with a plastic mask tightly covering my face to hold my head still, a mask that was, in both cases, custom-made for me from plastic mesh. With the CyberKnife, sophisticated software is aided by ongoing CT scanning that tracks the tumors' positions and reacts to even the tiniest movements of my head. The CyberKnife shoots beams of high-dose radiation into the tumors from multiple directions using a high-energy x-ray machine mounted on a robotic arm. Painless and noninvasive, there's no cutting involved despite its name. But extreme accuracy is essential so that healthy tissue is spared while the tumors are destroyed. Relying on precise calculations and a great deal of planning, any targeted radiation procedure (either CyberKnife or SRS) involves an impressive team: physicists (like my sister, Maria Czerminska, an oncology physicist in Boston), radiation oncologists (like Dr. Collins at Georgetown and Dr. Aizer at the Brigham), and dosimetrists, who calculate the doses of radiation and determine the optimal beam trajectories to minimize damage to the healthy brain tissue.

As the CyberKnife zaps my tumors, I lie as still as possible and gaze at the ceiling in the darkened room. My mind wanders into meadows and woods; I visualize a brightly shining sun and follow imaginary kites that float in a blue sky. In my head, I rhyme verses in Polish about the wounds in my brain becoming filled with green grass and blooming violets while the sadness of these many days of stress slips slowly out into the woods.

BRAIN RADIATION

U naszej mamy glowa dziurawa,
W dziurach wyrasta zielona trawa
Przez dziury smutek splywa do lasu,
Na smutki szkoda jest glowie czasu.
W zielonej trawie fiolki, dmuchawce
I porzucone stare latawce.

W dziurawej glowie figle, chichotki.
Lekka jest glowa zielonej trzpiotki.

———

Holes like winter potholes dot my broken head,
They quickly fill with mud and make a flower bed.
Green grass covers the holes, flowers begin to bloom,
Dandelions and violets take over the holy gloom.
Sorrow like water seeps through my ailing brain

It soothes the worried soul, eases the gnawing pain.
There is no point to worry having a grassy head.
It's funny,
It giggles and chuckles,
It laughs and dreams,
It's not dead.

Finally, I am sent home, tired and stiff but relieved that one more mission in my battle to live has been accomplished. For a little while, there is nothing to do but wait and hope.

I take it easy the next day. Surrounded by my husband and children—Witek and Cheyenne are here now too—I feel almost joyous, as if we are back to a normal life.

The following morning, less than two days after the CyberKnife procedure, I wake up early feeling healthy and strong and as if nothing extraordinary has happened in the previous weeks. It's a beautiful summer day, and I propose we go for a morning of light exercise in the woods of our favorite training ground, Prince William Forest Park. A gorgeous expanse with miles of hiking and running trails, the park was developed by the Works Progress Administration during the Great Depression.

This month, Witek, Cheyenne, and Kasia are preparing for triathlons. I was forced to put aside my own triathlon training after getting sick in January, although throughout this whole ordeal, I've never stopped working out—almost every day, no matter how I felt, I ran, walked, swam, or cycled. Today, as always, I'm eager for some physical activity. Even at the not-so-fast pace I now take, walking in the woods relaxes me. With my loved ones around me, it will be a great escape from the doctors and hospital rooms. I crave this trip. I need to feel as normal as possible.

The asphalt road through Prince William Forest Park is a loop of about seven and a half miles of hilly terrain. Whenever I'm training for a triathlon, I cycle it four or five times and then run it once. Since I have just been released from the hospital after being treated for profound brain swelling and since I'm two days post-radiation, I decide to take it easy; I will walk just one seven-plus-mile loop.

"Are you sure?" Mirek asks, concerned.

For our entire marriage, we've always checked on each other, but since I've gotten sick, Mirek has been obsessed with my safety.

"I'm fine, absolutely fine," I assure him.

Mirek loads his bike and Kasia's into our Toyota RAV4, and we head out, Witek and Cheyenne following in their car. It is already hot when we pull into the small parking lot we always use. We agree to meet back at the car when our workouts are finished and reward ourselves with a picnic lunch in the park. They all volunteer to check on me as we pass one another along the road.

Witek, Cheyenne, and Kasia take off on their bikes. Mirek kisses me on the cheek, hugs me, and pedals off on his own.

I hit the asphalt road and begin to walk, my arms swinging purposefully, my stride long and determined. The smell of the woods, the chirping birds, the waving branches of the tall trees make me feel free and happy. I breathe in deeply, and my lungs fill with fragrant air.

After about an hour, I pass a wide field that is yellow with chanterelle mushrooms. Chanterelles—meaty golden mushrooms with exotic ribbed underbellies and a strong peppery smell and taste—are a family favorite. They evoke memories of Poland, where they were abundant and we picked them in the woods around our summer home or on the outskirts of Warsaw. We love to cook them in various sauces or sauté them in olive oil and serve with scrambled eggs.

I'm delighted to see so many chanterelles and eager to gather as many as I can. But I have no bag to put them in so I keep power-walking. Luckily, Mirek soon rides up on his bike.

"There's a field full of chanterelles just a little ways back," I tell him. "Can you go to the car and get a bag and pick some? We'll have them for breakfast tomorrow with scrambled eggs."

Mirek takes off and I continue on my walk. After another ninety minutes of vigorous walking, I finish the seven and a half miles and reach the parking lot.

Although I was filled with energy when I began, two and a half hours ago, I'm now completely exhausted, so physically and emotionally spent that I feel like I've run a marathon. With an almost primal urgency, I feel a desperate need to rest and eat immediately.

But to my surprise, Mirek isn't back yet.

I will call him.

But—I can't remember his number. And for some weird reason, I can't recall how to find his number on my cell. I fumble with the phone. And then I forget what I'm doing.

What's this about? Oh, yes, I'm trying to call Mirek. But where's his number? How do I call him?

I fiddle with the phone, trying to figure it out. And over and over, I have to keep reminding myself what I'm trying to accomplish. Finally, I locate his number in my contacts list and call him.

"There are so many chanterelles!" he says excitedly. "I've collected a huge bagful."

"We need to eat lunch right now," I say angrily.

"Great!" he says. "I'll be waiting here for you."

"No, no! *You* come *here!*"

"I can't bring them on the bike, they'll get squashed," he answers. "I'll wait for you on the side of the road."

Only after we hang up do I realize I have absolutely no idea how to find him.

I know this park like my own backyard. I've cycled and run and walked it scores of times over the years. Just an hour or so earlier, I'd told Mirek where to find the chanterelles. But now my mind is frozen. I can't picture his location at all. Driving the car to him seems like an insurmountable challenge, a feat completely beyond my ability.

I stand with cell phone in hand, fuming. *How the hell am I going to find him?*

I decide to call him back. But once again, I can't figure out how to find his number.

I can't seem to think straight. *How do I call him?* I concentrate very hard and try again. And again. After enormous effort I find his number, but by now I'm very frustrated and growing angrier.

"Come here *now*, Mirek!" I snap. "I don't know where you are!"

"Just drive down the road," he responds. "You can't miss me."

"Which way?" I plead.

"It's a one-way road, darling," he says.

That confuses me even more. What does *one-way* mean? It makes no sense. Even though I'd driven on this road dozens and dozens of times, getting to Mirek seems like a highly complex puzzle that I cannot solve.

"I don't know where you are!" I repeat, my voice rising.

"It's a loop. Just drive down the road," he says, and he hangs up.

I stay in place, seething. I look for his number to call him back; it takes even longer to find it this time.

"Where *are* you?" I ask, on the verge of crying.

"I told you!" he says. "Just get in the car and pick me up."

"No, no, you come back. I'm tired!"

"It will be much faster if you drive here," he says, beginning to get angry himself.

By now, Cheyenne has finished her run and arrived at the parking lot. She listens with a quizzical expression as I argue with Mirek. When I tell her why I am so upset—"I don't know where he is!" I whine—she offers to go pick him up in her car.

"No!" I snap. "Leave him there with those stupid chanterelles."

"Why don't we take a little walk," she offers gently. "Just until Witek arrives."

But I do not want to walk with her. I am furious. I decide to go find Mirek myself. I get into the car and start the engine. But should I turn left or right? What did Mirek mean by a loop? I can't picture it in my mind.

Finally, I pick a direction randomly and head down the road.

I'm bewildered and increasingly irritated. The trees, the fields, all of them look familiar yet at the same time unrecognizable. And no matter how hard I try to retrieve the concept of a loop from the depths of my brain, it eludes me.

I drive very slowly, growing angrier and more exasperated. I begin to fixate on Mirek's behavior.

I'm tired, I need to eat now, and he wants me to find him? He might as well be lost in a giant forest in a foreign country. This is Mirek's fault, entirely his fault—he gave me the wrong directions!

Up ahead, I see Kasia and Witek running along the road in my direction after finishing the cycling portion of their workout. Unlike every other time in my life, the sight of my beloved children does not make me happy. I stop, and Kasia gets into the car. Witek keeps running to meet up with Cheyenne at the parking lot.

Seeing my scowl, Kasia asks, "Mom, why are you so angry?"

"Mirek is taking so long! I want to go home! *Pieprzone kurki!* Damn mushrooms!"

"Mirek's picking mushrooms," she says soothingly. "We'll be there soon." She gives me very simple directions—"Just keep going straight, Mom"—but I'm angry with her too.

"How can you be so sure that I need to go straight?" I say. "It's annoying. Why am I at fault for this stupid loop and the park and everything?"

Her eyes well with tears. "We're all here for you," she says. "Why are you so angry?" she asks again.

"Because he's late!" I half scream.

But then, ahead of us, there is Mirek, standing at the side of the road smiling and waving, his bike against a tree, a full bag of mushrooms in his hand. He puts his bike on the car and climbs in with his bounty. At first, he doesn't notice my dark mood.

"Look at all of these!" he says happily.

I refuse to look. I want to toss the chanterelles out the window.

"I need to eat!" I shout. Mirek looks at me with a shocked expression.

Kasia offers to drive, and I move into the passenger seat, too tired to argue with her. I sit in stony silence as we head to our picnic spot, where Witek and Cheyenne join us. As they lay out the tablecloth and unpack sandwiches, fruit, and granola bars, I fume. We eat quickly, with little conversation, my inexplicable anger unnerving them. The food helps a little, but I'm still incredibly tired and angry at the entire world.

When we get home, Witek washes the chanterelles and I go upstairs for a nap.

An hour later, I wake up and head to the kitchen to begin making dinner. Each day, cooking has become harder for me, but now, as I stand here, I can't remember what to do. Not even the simplest steps.

"Where are the pots? Where are the spoons?" I grumble. "Why can't I find anything?"

Everything is gone! My family has gone behind my back and rearranged my kitchen! I slam drawers and pull open cabinets in a frenzy. *It's all wrong, everything has changed. Why would they do this to me?*

I finally find what I need. But as I set to work, the simple recipe that I've made hundreds of times seems like a complex mathematical equation.

I try to recall the ingredients and locate them in the pantry. But it's so hard! I grow more agitated, swearing and banging cabinet doors. Mirek peeks in and offers to help.

Kasia and I relish a breakfast of chanterelles and scrambled eggs the day after our fateful excursion.

"No!" I shout. "I make the dinner! I always make the dinner! I'm not going to stop now just because you've moved everything around!"

I manage to create some strange concoction that they politely eat at a tense dinner table where no one talks. For the rest of the evening, I barely speak, and when I do, it is only to criticize them.

Despite the fact that I am clearly struggling with so many simple tasks, I continue to be very driven, especially with regard to getting back to the sports I love. I have a powerful desire not to interrupt my training routine and daily life. Altering my habits means I have to accept that I am not okay. Conversely, willing myself through a draining workout proves that I can overcome any obstacle and beat any foe—even brain cancer.

But my feelings of power and strength are an illusion, born in large part of the high doses of steroids I am taking combined with my natural determination to survive.

And while I'm feeling better, my frontal cortex is not functioning normally. Just days earlier, it was being squeezed and pressed against the inside of my skull because of the inflammation and swelling in my brain. If I hadn't been treated with high doses of steroids in the ER, I might have suffered permanent damage to my frontal lobe. I could have lost forever critical cognitive functions, such as judgment, as well as my social skills, empathy, and personality. In fact, if the inflammation and swelling had not been caught in time—had my brain stem been choked off and sent me into cardiopulmonary arrest—I could have died.

Since my frontal-lobe function is still compromised, my brain cannot respond appropriately when faced with complicated or demanding tasks. That morning before we went to the park, when I was still at home in a quiet, familiar environment, I behaved normally. For that reason, there was a strong emotional incentive for all of us to believe I really was fine, especially when I insisted I would have no problem walking in the woods.

But after walking for seven and a half miles, I became exceptionally tired and hungry, and at the end of the two and a half hours, my brain was in no shape to handle much of anything. Depleted and exhausted, it had entered survival mode. When called upon to do anything that was even slightly complicated—such as finding Mirek's phone number, calling him, processing his request to locate him, retrieving from my memory a visualization of the road, understanding it was a loop, or recalling which way the one-way traffic went—my traumatized brain fell apart. With that kind of information overload, the neural connections within and between my frontal lobe and other brain areas became clogged, like a traffic jam in my head. Finally, my advanced thinking came to a near-complete standstill. Sensing itself endangered—too much going on, too many demands!—my brain ignored everything but primal necessities. *Rest, rest, rest!* it told me. *Rest and eat! Do not engage in anything else! Your survival is in jeopardy!*

Try asking a hungry toddler, or even a hungry eight-year-old, to solve a puzzle after you explain that dinner will be ready soon. She will throw a tantrum, kick and scream, call you names. Since her undeveloped frontal lobes will not mature until her mid- or late twenties, she is controlled in large part by her instincts and basic emotions related to survival. She has no impulse control, is not rational, has a short attention span, and cannot understand the concept of waiting for a reward—food—that will come later. Her brain is telling her one thing: she must eat now.

Try the same experiment with a marathon runner who is just reaching the finish line. He would rather slap you in the face than try to solve a simple algebra problem. With its energy stores almost completely depleted, his brain is hoarding whatever's left for the region essential to survival: the primitive limbic brain, which operates autonomous functions, like keeping his heart and lungs operating, and regulates basic emotions like fear. His brain has switched off the luxurious, sophisticated frontal lobe, which enables problem-solving and other higher cognitive functions that make us human, including the ability to evaluate options in order to make judgment calls. For the exhausted marathon runner, these more refined skills simply aren't as essential as the basic brain functions that will keep him alive, so they go into a kind of hibernation until there is enough energy for them to spring back.

I experienced this phenomenon myself when I ran marathons. In the final several miles, I could never calculate my pace because my brain could not do the necessary arithmetic. As I approached the end and focused only on getting across the finish line, I was in a zombie-like mental space. If anyone interrupted my singular focus, I reacted with anger. When my husband tried to tell me encouragingly that the finish line was close, I would snap: "Bullshit! It isn't close enough!"

Or take my elderly mother. Otherwise brilliant and fully functional, she cannot engage in more than one task at a time these days because her frontal cortex—which deteriorates as a person ages—gets overloaded easily. When too much is happening around her, she becomes disoriented, panicky, and angry.

Similarly, patients with schizophrenia don't perform well under conditions of increased cognitive pressure. Brain-imaging scans show that when people with schizophrenia are presented with overly demanding tasks, such as solving complex tests, their prefrontal cortices are not activated at the same level as neurotypical people's. When too much is asked of them or there is too much stimulation in the environment, their already compromised brains fall apart further. They might behave angrily and inappropriately, like I did during the park misadventure.

Before we went to the park, I was largely fine. But with too many demands on my brain that day, the most advanced part of it—the most human part—simply shut down. My meltdown was clear evidence that I was not out of the woods yet. And I would need even more aggressive treatment to stay alive.

9

What Happened, Miss Simone?

*O*ne afternoon at the beginning of July, I'm walking with Witek on a quiet, empty street, holding tight to him as if I'm afraid I'll lose him. We're headed from home to a nearby pharmacy to pick up my prescription for oral steroids. I've been having such trouble with directions lately that we're going hand in hand, my son and I.

I look at the contours of his slim face and his strong muscular body. Witek is all I could have wished him to be—a scientist working on the brain, an athlete, and a kind man. He finished his first Ironman just a couple of weeks ago, while I was in the ER, and now he's training for the next one. His goal is to qualify for the Kona Ironman in Hawaii, the crown of the Ironman competitions. He's found the love of his life, Cheyenne, who shares his fascination with endurance sports. I'm proud of him and glad he is by my side.

But today I feel acutely that our lifelong roles have reversed. I am no longer his strong mother-protector. Instead, he's leading me as if I were his little daughter. His presence gives me a warm sense of safety but I feel odd—fragile and dependent.

We talk about everyday things: his work, friends, the weather. The air is damp, the sidewalk wet. As often happens here in July, there have been some severe storms. But I don't remember them. I know there've been storms only because I see tree limbs scattered all over our neighborhood, and several houses have been damaged by huge branches falling on their roofs.

We pass a car with half of a tree splayed across it. The car is crumpled, the metal mangled. The windows are shattered, the glass all over the sidewalk.

"Look at this car!" I say to Witek. "What a terrible thing. Wow, half the tree fell on it!"

"Yeah, bad luck," he agrees. We continue walking.

Inside the pharmacy, I cling to Witek, not confident enough to let him out of my sight. But as we wait for the prescription, he wanders off, checking the products on the shelves.

I find myself uneasy. There are too many people here, too many things going on. I start to meander but have trouble navigating the store. I knock into shelves, bump into other shoppers. It's as if I've lost my balance or can't estimate my distance from objects. I don't quite sense the boundaries of my own body, can't really feel where it starts and ends, have no real awareness that *this* is me and *that* is the outside world. I feel as if I've fused with my environment.

I'm frightened. *Where is my son?*

Witek finds me, my prescription in his hand, and we head for home, walking slowly as I hold on to his arm. We pass a car with half of a tree splayed across it. The car is crumpled, the metal mangled. The windows are shattered, the glass all over the sidewalk. There must have been a storm last night.

"Look at this car, Witek!" I say. "What a terrible thing—a tree fell on it."

Witek gives me a strange look. He seems surprised and uneasy. I don't like it.

Something is wrong. What have I done?

As I look into his face, I grab him tighter. I'm afraid to let go.

Like a person with early-stage Alzheimer's or any number of other mental conditions, including brain injuries, I am losing my short-term memory. While I retain sharp memories of my childhood and various long-gone incidents—which is why I can write so much about them—I can't remember what happened just minutes ago. Short-term memories are processed differently in the brain than long-term memories, so people with dementia can often remember events that happened in their childhoods but have trouble recollecting what they had for breakfast that day. Long-term memories are tucked away in our brains for safekeeping with strong emotional attachments tied to them, since they may be useful for survival. Short-term memories appear to be more like temporary factoids waiting to be categorized and evaluated. If important, they'll be stored. If they're unimportant, they're not tagged for retention and will vanish.

But I don't realize that my memory is faltering. I don't realize that I'm missing anything at all.

"Mom, we saw that car on the way to the store," Witek says carefully. "You don't remember?"

I'm not sure. I'm not sure of anything anymore.

Late the next morning, Mirek and I drive to a nearby trail that snakes through the woods behind the houses in our neighborhood. Strolling among the trees, we move slowly, holding hands. We talk about what to make for dinner, what shopping needs to be done—the small talk of everyday life. But mostly, we enjoy the silence.

Mirek decides it is time to head back. In less than half an hour, we reach our car, which is parked on the side of a quiet street. When he gets into the car, I tell him I'm not ready to stop walking. I like to move. I can rarely stay still—at the office, I've always been one to jump up and stretch and walk through the labs, check up on things—and I try to find every opportunity for more time outdoors.

"I will walk home," I say. "I need the vigorous movement, okay?"

He hesitates, then tells me he's not sure I'll be able to find my way back.

"Oh, please, we're just a mile away! Of course I can get home," I say. "I know these streets as well as you do."

I turn and start walking, fast. Moments later, he passes me in the car. I wave and he waves back, smiling.

It is a hot and hazy July afternoon. The world around me is quiet, which I cherish. A few birds chirp cheerily; cars hum in the distance. I walk happily, legs moving briskly, arms pumping to increase circulation in my upper body.

At first, my stride is very fast—but not for long. I soon tire and slow down. My body is a shadow of what it was before the treatment and stress of illness took their toll. I've lost a lot of muscle mass, the result of high doses of steroids. I look down at my thighs, once muscular and strong, capable of running and cycling for tens of miles over rocky paths, desert sand, snow, all kinds of terrain in all kinds of weather. Now I see scrawny legs barely able to hold me upright.

I continue walking nevertheless, convincing myself that I will fight through this illness, improve my pitiful form, and return to my athletic self.

I pass intersection after intersection, dutifully checking the street signs. I'm being careful; I don't want to get lost. But after a few hundred yards, I stop recognizing the streets. I know their names, all right—that is, they sound familiar—but I can't pull up from memory where they lead or which direction they run.

Okay, I know I am a little more than a mile from my house, so it cannot be very difficult to find, I assure myself. I keep walking.

I can't be lost—not when I'm so close to home. I just need a bit more time to come upon the correct street and recognize the houses. Then I'll easily find mine.

I don't panic. I don't even worry. I just walk and walk. All the quiet houses look exactly the same; all the deserted streets seem identical. There is not a single soul outdoors; the heat must be keeping my neighbors inside. No one's mowing the lawn or trimming hedges. There's no one to ask for directions.

I keep walking. But I'm getting really tired. And I need a bathroom. I really need to pee.

I know there are no public bathrooms for miles, and no woods here either, just house after house after house. I look around to find some bushes where I can do my business. But there are none to suit the purpose. Just manicured lawns and evenly mowed grass and nicely trimmed trees.

I can't wait much longer.

I can't wait at all.

I pee. I pee in my shorts. I don't pause or even slow my pace. I walk and pee. I don't want it to happen, but it does, as if it has a mind of its own. I'm not concerned that somebody will see. I stroll along the street past the houses of my neighbors, peeing my pants like a little child, and I don't care.

An hour or so later, I flag down a car at an intersection and ask the driver for directions. But I have trouble explaining where I want to go. I tell him the address but he doesn't know where my street is. In an effort to figure out where I live, he asks additional questions, like whether I'm near this or that neighborhood landmark, but I'm unable to give any meaningful details. He offers to drive me around but I refuse to get in his car. It's not that I'm afraid to ride with a stranger; it's that I want to walk. That was my plan, and nothing is going to change it. He volunteers instead to lead me to the closest major road in the hope that it will jog my memory.

My stride is unsteady as I walk along following the man's car. But

I'm not troubled by my wet shorts. He drives very slowly, and I tail him through the monotony of the red-brick façades of the houses in our northern Virginia suburb. When we reach the road, the pieces of the puzzle suddenly fall into place. I recognize my neighborhood: that little house with yellow siding on the corner, the brick mansion across the street. I now know that I have to turn left onto the busy road, and a hundred yards later, turn left again. I see my house.

Mirek greets me with relief. He cannot understand why I took so long.

"I got a bit lost," I say. "These streets weave in and out, and it's hard to follow them."

"Okay," he says, giving me a kiss. He's clearly happy I'm home.

"And I wanted to pee so much that I wet my pants," I say.

He looks down at my wet shorts and legs. "Oh, dear," he says with affection. "Just wash it off."

This incident marks the first time in my conscious life that I am incontinent. For the next month or two, I will sometimes have trouble controlling the reflex to urinate (in medical jargon, *micturate*) that arises in response to an increase in pressure in my bladder. If I get stuck in traffic on my way to work, as soon as I park my car on campus, I have to run to the nearest building to find the bathroom.

Is it possible that the inability to inhibit the urge to urinate has something to do with brain function? As it turns out, it may be linked to dysfunction of the medial surface of the frontal lobe, which is the cortical center of micturition. A majority of stroke patients with frontal-lobe lesions develop urinary incontinence, and patients with frontal-lobe tumors often can't tell their bladders are full until the very last moment, when they can no longer control the need to pee. Incontinence is also a common problem in people with dementia, and, in general, it's a common disorder in older people. There can be a number of reasons for it, many of which have nothing to do with brain disease, such as a urinary tract infection, inflammation of the bladder wall, or prostate problems. But when someone my age suddenly becomes incontinent, it can be a sign that the brain is at fault.

The inability to control urination may be a symptom of other mental illnesses besides dementia. A former colleague of mine at the NIMH, Dr. Thomas Hyde, a neurologist and a schizophrenia researcher, hypothesized that children who would go on to develop schizophrenia took longer to learn to control their bladders than children who did not later develop the disorder. And indeed, the research found that adult patients with schizophrenia had had a higher rate of urinary incontinence in childhood than their healthy siblings. The impaired bladder control that many patients with schizophrenia experienced as children might be connected to delayed maturation of the prefrontal cortex, he believes.

It's yet another irony for me. While I don't have schizophrenia, I am living through some of the processes of a disease that I've spent my life studying and trying to cure.

Throughout my life, I've been quick to react, independent, confident, and stubborn. But now, these qualities are reaching an absurd level. I'm in a constant hurry, skipping mindlessly from one activity to another. My attention span is completely shot. When I try to read, I go faster and faster over the words but have little idea what I've read. I jump from page to page, story to story, sentence to sentence, word to word, but I can't absorb their meaning. I continue to talk on the phone to my children and sister every day but I don't finish a single conversation. I cut each of them off midsentence and run somewhere to do something of great importance although I'm unsure what it's supposed to be. I feel anxious and stressed out but I don't know why. And I don't listen to what Kasia and Mirek and Witek try to tell me. I know best. They don't know nearly as much as I do!

One day, I read a story in the *Washington Post* about a student from a nearby high school who thought she'd been accepted to several Ivy League schools but found that the schools had misled her. I describe the story to Mirek, but when I finish telling him what I've just read, he gives me a strange look.

"That's not what happened at all," he says gently.

"I just read it!" I insist. "Don't you think I know what I read?"

"You got it backwards," he says. "She claimed Harvard and Stanford both wanted her but it turns out she made it all up."

"No, no. You're completely wrong, Mirek," I say angrily, but he gives me a sad smile.

With each new day, I feel increasingly confused. The world seems to be whirling faster and faster all around me. I have trouble catching up with it. I don't understand what's happening, can't follow its meanings. It races forward as I'm left behind.

In early July, the newspaper announces the grand opening of the brand-new Giant grocery store that I've waited so long to enjoy. I never thought I would live to see it.

The Giant has taken on a strange significance for me. It epitomizes the cruel passing of time and uncertainty of my own existence, the fragility of my life despite my physical strength, my athletic ability, and my stubborn optimism. Indeed, as I have endured my illness, I've begun to resent the massive concrete construction.

That stupid store will be standing there when I am gone.

Now that I've lasted long enough to witness the grand opening, it's really important to me to go. We all decide—Mirek and I, as well as Witek, Cheyenne, and Maria, all of them here visiting me—to head there for the festivities. But the moment we park and I open the car door, I recoil. I'm repulsed by the large crowds and the loud music of a live jazz band inside the front entrance that is welcoming shoppers. My family doesn't notice my reaction. Witek, Cheyenne, Maria, and Mirek are thrilled. We've all loved jazz for as long as I can remember. They stand and watch.

I'm fuming. Under my breath, I mutter, "What the heck! Why on earth is the music so loud? I can't even communicate with my own family!"

They don't see how much I hate this. I begin shouting over the music. "This is horrible!" I yell. "It's too loud!"

They look stunned, and they try to calm me down.

"Mom, this is nice," says Witek. "These guys are great." Witek plays clarinet and guitar, and he learned flute in Hawaii when he spent a year there managing a coffee plantation. I like it when Witek plays; it soothes my soul, relieves my moodiness. But this jazz is hurting my ears, pounding deep holes inside me like a jackhammer. It's painful.

I bolt from them and run through the store to search for the main office; my family races after me. As Witek and the others try to stop me, I demand to see the manager.

When the manager appears, I shout: "Stop the music! It's too loud! It hurts my ears! Stop the music!"

She looks at me, then at my family. Before she can respond, I turn and storm out.

I rush past the band, and the music causes me physical pain. The notes are like knives stabbing my body.

My family catches up with me, and as soon as we climb in the car and close the doors, I feel better. It's much quieter, and we drive home in silence. I am calmer already.

"What a band that was!" I try to joke.

No one responds.

My hypervigilance—my body constantly on high alert, and the sense I have that I'm participating in every event with my whole being—is possibly being triggered by stress or anxiety. That anxiety, in turn, gives rise to more stress and anxiety. Making it worse, I have the vague feeling that I'm not in control of myself or the world around me anymore. That loss of control makes me angry.

My extreme reaction to sensory overload is common in people with brain trauma, autism, and many other brain conditions. Normally, the brain is able to sort through the sensory information that comes at it and prioritize what's important and what can be ignored. When this filter mechanism doesn't work, the brain can become overwhelmed by all the information it's trying to process, like a computer bombarded by too much data. The brain can no longer distinguish between what it's safe to ignore, like the sounds of distant traffic or the sensation of wind on your face as you walk along, versus what is important, like the honking of the car that's about to hit you. This horrible jumble of noises and sights and smells can be very upsetting. When faced with significant sensory overload, some people have a reaction akin to a panic attack, like what I experienced at the supermarket.

In my altered state, I couldn't even begin to comprehend what was happening to me. And scientists are still far away from fully understanding the mechanisms responsible for anxiety, responses to stress, and attention. We do know that they are disrupted in certain mental disorders, including ADHD (attention deficit hyperactivity disorder) and PTSD (posttraumatic stress disorder). We also know that a complex network of neuronal connections between many regions of the brain must operate properly in order to guide a person successfully through the jungle of human experience, which presents all kinds of stressors.

In my damaged brain, even the most innocent of stimuli, something as pleasant as a jazz band, is too much. I can't handle it.

That night, Mirek and I are watching a movie on the huge flat-screen TV in our basement turned home theater. We snuggle on a comfortable leather couch we bought six years ago when I was getting chemotherapy for breast cancer. We lie so close that we feel each other's hearts beating, our lungs drawing air, our warm bodies intertwined. Mirek holds me tight, caresses my arm, gently tickles my hand.

I feel safe cuddled next to him like this with his warm, loving hand on mine. But inside my head, a strange and not entirely unpleasant chaos is stirring.

Black and white—death and life—white and black—life and death —black—black—black.

We're watching a documentary about Nina Simone, *What Happened, Miss Simone?* The images are flying by . . . the music is blasting . . . her deep, strong voice is entrancing. I'm hypnotized. I can't move. I'm experiencing this with my whole body. Her voice, her overwhelming persona penetrate me not only through my eyes and ears but through my skin, flooding me with emotions, shaking my insides. I'm mesmerized. I quiver as if I'm absorbing too much for my battered head to consume.

"Too loud for you?" Mirek asks. "I can turn it down a bit."

"No, no, please! I love it!" I say.

Black and white—white and black—black, black, black.

The images on the screen flash like a monochromatic kaleidoscope, sharp edges, multiple reflections, fast, fast, fast. While it's hard for me to follow the story on the screen, I can't tear myself away from what I'm seeing. Simone is beautiful, phenomenal, strong and brittle at the same time, her life passionate, dark, and tragic. I cling to Mirek for support and think about my own impending death.

Black and white, black—black—black.

"Can you pause it for a moment?" I say.

I leap up and race out of the basement and up two flights to my office. I pull open the bottom drawer of my desk and frantically sift through a pile of documents.

There! Found it!

My health directive. I must add something to it, right now. Fast, before it's too late. *Do not resuscitate.* I must add those instructions immediately.

I search for a pen and scour the paperwork. Where to add the words? I struggle as I try to read it. *Here, I'll put it here.* I try to write but can't remember how to spell *resuscitate*. My handwriting is shaky and hard to read. The letters I write are squirming, wiggling. They don't look like English or Polish or any other recognizable language.

I'm terrified I won't be able to convey my desperate desire: *Do not mess with my body, do not traumatize it, be gentle and leave me alone when the time comes and death is near. Don't be brutal. Don't force me to live when my body quits.*

I scribble something that's supposed to be *DNR* on my health directive and run out of my office. I need to be back in Mirek's warm embrace. We've been such an excellent team throughout the years: through my divorce and the death of my ex-husband, through raising the kids in a strange country and buying and renovating our home when we had very little money, through my breast cancer. And now, through this illness, which looks like it's going to be the last, most difficult passage in our lives.

I run downstairs, skipping steps, feeling ready. But ready for what? Ready to lie down next to Mirek and embrace? Ready to die? Both? I push away that grim thought. I have edited my health directive. I've done something constructive, and I can rest.

10

The Light Gets In

The summer of 2015 continues to torture me and the world around us. The unrelenting heat is killing the grass; flowers are wilting and dying.

One particularly sweltering day, I open the door and a blast of hazy, hot air hits me in the face as if I've opened a giant oven that could kill me. But I am not ready to die. I slam the door shut and retreat inside to my cool nest with its air conditioning that hums day and night. Since my doctors don't want me to drive, I spend most of my days sitting with my laptop on a couch in the living room, taking care of brain-bank tasks or writing my memories.

The steroids I'm on are decreasing my brain's inflammation. However, they're taking a hefty toll on my body. My normally slim, long face rounds up like the moon, which is typical for patients on steroids. My body shape changes too, dramatically and so fast it is terrifying to watch. Within weeks, my muscles and athletic figure are gone. My body becomes heavy and inflexible. I look down with dread at what used to be my cyclist's thighs and runner's calves and don't recognize them—they're emaciated and weak. My flabby belly sticks out no matter how hard I try to suck it in. My swimmer's muscles, of which I was so proud—triceps, biceps, and latissimus dorsi, the large shoulder muscles—disappear completely, and the skin fills in with Jell-O-like fat. I gain another slab of fat on my upper back just below my neck, turning me into a kind of hunchback. I go from size 4 to size 8 in a matter of weeks. After that last session of radiation, I also start losing my hair. It's coming out in thick clumps. I hate to look in the mirror; I'm a bald, elderly caricature of my previous self. Am I still the same person? How much change does it take to completely obliterate my sense of me?

I continue to exercise but instead of running and cycling, I mostly walk in the nearby woods in the early mornings and late afternoons. I go shopping with Mirek, holding on to him tightly, afraid to get lost, afraid to fall. My legs are not holding me up as they should; my balance is off. The world around me is swaying, coming in and out of focus. I am not sure of the reason: Is it in my brain or body? Is it mental or physical? I cannot tell. The two are inseparable.

But I can write and work day and night without a break. Steroids fire me up, just as they did in January when I was recovering from brain surgery. Once again I am like a manic person, a woman possessed, a highly driven insomniac. Since I can't drive, I work from home. I hold long teleconferences with my colleagues, write reports, reply to e-mails, plan experiments, fill out administrative forms, make arrangements with morgues to collect brains for our studies. I can do these things, but it takes a lot of effort. I forget words and tasks. My own brain is still out of order—dotted with horrific craters, enveloped in clouds of inflammation. I swim in and out of the real world.

But as the days pass, I have more and more moments of clarity. I don't know what's happening inside my brain but the swelling must be receding because my mind is returning. I start to realize that I've been through something very strange, a bizarre and unusual odyssey. Slowly, I also begin to understand where that journey has taken me: into insanity, and now back.

As if from some previous life, as if from the deepest fog of perception, images of my recent past begin to emerge. I'm regaining my hold on everyday life and on reality. It's like I'm clawing my way up from a black hole and slowly beginning to recognize my surroundings and see the sun. And I'm starting to realize how deep that hole was.

I ask Mirek and the children about the past weeks, how I behaved, what I said, what was different about me. They aren't eager to talk. They share as little as possible. They are traumatized by my alien behaviors and the still-looming possibility of my death. And they are afraid that the pretender version of me—the mean one who criticizes them relentlessly, who is distant and unloving and confused and angry—might return.

But sometimes, they tiptoe into testing what I recall, to see whether I have any idea of what the past two months have been like for me—and for them. Witek brings up our walk to the pharmacy not long ago. "Do you remember it, Mom?" he says. "That you couldn't recognize that fallen tree that you'd just spotted half an hour earlier?"

At first I don't recall anything.

Was I even there? When did it happen? Was it really me?

I focus and close my eyes. I strain my brain and squeeze my eyes tighter, and as I do, I begin peeling away layer after layer of my own forgotten life. I can smell the wetness of the storm and visualize our stroll along the sidewalks strewn with branches and debris.

The motto that adorns the main hallway of the Georgetown University Hospital pops into my mind: *We are all broken, that's how the light gets in.* It speaks strongly to me, and I whisper to myself, "Through my broken brain, the light starts getting in."

The memories of the past two months begin creeping back. Like scared little critters who've been hiding in the corners of my mind, they begin to emerge, testing the ground first before peeking out cautiously from the folds of my beaten brain. With effort, I can recall the bare facts and see the things my family mentions: the tree branch, the sidewalk, the damaged car. I start recalling more events.

Strangely, however, I cannot fully resurrect the emotions of that time. It's much harder to recall how I reacted and what I was feeling. On the infrequent occasions when my family tells me about something odd that happened, I listen carefully but cannot connect the factual description to the turmoil they went through. I don't remember any of that. It's as if my emotional memory resides somewhere else, someplace to which I still have no access. Maybe the feelings were never encoded in my brain at all.

Mirek asks, "Remember that awful dinner after we picked you up from the hospital? You were breaking my heart with your empty eyes, frozen facial expression, harsh words. You were so mean, so cold."

I try hard to recall. I ask about the details—what I made for dinner that night, where we were sitting, who said what.

"Kasia and I got up from the table and went into the kitchen to cry. It was unbearable to see that you were not your normal self. We thought you were gone forever," Mirek says, and he chokes with emotion. "You reminded us of Kai, the little boy from Hans Christian Andersen's fairy tale 'The Snow Queen.'" My husband's eyes fill with tears.

I strain my brain again, and the images emerge as if from a movie I'd watched years before.

Yes, the dinner, I remember. I was cooking and it was not coming out as I expected. There was something weird about that dinner. But what? Was I distant and cold? Did they cry; were they sad? I don't remember. Maybe it happened to another me, another person entirely?

I do, however, remember the story of the little boy named Kai. The tale terrified me when I read it as a child. The story goes that two children, Kai and Gerda, live in fairy-tale happiness until a mean goblin with the power to turn beauty into ugliness breaks his magic mirror. Hundreds of millions of pieces scatter around the world. One splinter pierces Kai's heart; another one gets into his eye. His heart turns into a lump of ice; his eyes see only evil. Kai becomes cruel and aggressive. He abandons Gerda and his loving family and chooses to live in the eternal winter of the glacial palace of the Snow Queen.

A nasty goblin must have embedded a shard in my brain and made me insensitive to the ones I love. He turned me into a callous and unloving caricature of myself.

Now my icy heart is thawing and I am crawling back to life, one dreamlike recollection at a time.

How is it that these memories are returning?

The brain has a remarkable ability to heal itself after various kinds of injuries and assaults, a capacity that amazes scientists and doctors. Even patients with severe brain damage can sometimes recover nearly fully. While it's clear that excellent medical care and therapy can assist in recovery from a brain injury, how the healing process works remains unknown. The BRAIN initiative that President Obama launched in 2013 is aimed at revolutionizing our understanding of the brain, including how it recovers from injury and disease. But at this point, frankly, the brain's ability to repair itself seems nothing short of miraculous.

Unlike cells in other parts of the body, which constantly replace themselves, neurons in the brain do not as a rule regenerate. Experiments in mice have shown that a limited number of new neurons may grow in the hippocampus, the part of the brain that stores memories and one of the first regions affected by Alzheimer's disease. But it's probably an insignificant number of neurons, and it's unclear whether they ever become fully functional. It's also unknown if the same phenomenon occurs in the human hippocampus. We do know that in the brain regions critical for thinking, such as the prefrontal cortex, the neurons that emerged in infancy and probably even before remain the same throughout a person's life.

The fact that we retain the same neurons from the beginning to the end of our lives may be one reason we can consider ourselves "ourselves." What may change, however, are the connections between cells and among brain regions. Some connections grow stronger; some wither; some are damaged. If a region of the brain becomes impaired, new connections between cells may grow and help us recover some or most of the disabled function. But does that modify who we are?

I'm always surprised how little we change throughout our lives, even after traumatic experiences and serious illnesses. I was kind of myself—some version of myself—even when one-third of my brain was dramatically swollen. I am still myself now, as I continue to recover. But the tumors, the radiation, the brain swelling—all of these may leave their marks on my brain and my personality. They may lead to scarring, which can result in lingering damage to the brain. People who've received radiation to the brain or chemotherapy or immunotherapy may have ongoing cognitive problems, including memory issues.

When someone asks me how I'm doing—meaning, is my brain operating as it used to—I say, "I'm functioning just like before." But is it true? My attention span often seems shorter, and I get tired more easily. It's harder for me to concentrate. I cannot run or swim or cycle as fast, and my balance is not as good as it used to be. When I ask my family if I've changed and how, they say they're unsure. But it's clear this ordeal has affected all of us. There's no question that it has aged me. It has aged them too.

Despite the relief I get from steroids easing my brain's swelling, and the fact that radiation is killing the visible tumors, my family and I are acutely aware that there are melanoma cells lurking in my body. New tumors will most likely grow, and probably soon. They will spread wildly, out of control, and take over my brain like weeds invading a neat flower bed. Although I've received a kitchen sink full of therapy

—radiation and a combination of two immunotherapy drugs—I need more treatment, perhaps a whole bathtub of it.

So Dr. Atkins is adding targeted therapy, a final option that was floated at the very beginning of my treatment. Although there are some novel drugs in the research pipeline that I'm hearing about, at present, targeted therapy is the only option left for me to try. He says that I should be treated right away with a combination of trametinib and dabrafenib, two drugs newly designed specifically for a mutated gene involved in melanoma, *BRAF*. Trametinib inhibits the MEK1 and MEK2 proteins, and dabrafenib inhibits the BRAF protein. All three act in the same cell-signaling pathway, which becomes overly stimulated in melanoma cells and leads to their uncontrollable growth and proliferation. Two mutations, called *BRAF V600E* and *BRAF V600K*, account for over 95 percent of mutations in the *BRAF* gene found in melanoma patients. If a patient has no mutation in the *BRAF* gene, he's carrying what's called *BRAF* wild type and will not benefit from these drugs because the pathway in which they act is not abnormally overactivated by the faulty BRAF.

My tumor was tested genetically in March 2015, shortly after it was removed from my occipital cortex, and found to have a rare mutation, *BRAF A598T*, which occurs in less than 5 percent of melanoma tumors. In the genome, it is situated very close to the more common mutations, so it's possible that this gene makes a faulty BRAF protein just like they do. But nobody knows for sure. If my mutation does behave like the common mutations, then the BRAF/MEK1/MEK2–inhibitor drugs may be able to block the haywire activation of my melanoma cells and halt their proliferation. In any case, that's the plan. We're hoping the combination of both drugs will finish off my cancer.

These new drugs seem to be my last chance for survival. They are small molecules that can easily cross the hard-to-penetrate blood-brain barrier and get into the brain. By contrast, the antibodies employed in immunotherapies are large proteins, which, if taken orally, are quickly digested like all other protein products that we eat.

That's why they have to be given as infusions delivered directly into the bloodstream. Immunotherapy drugs don't actually get into the brain; instead, they modify immune cells (T cells), which can reach the brain. Trametinib and dabrafenib come in the form of very innocent-looking pills, which is much more convenient than an infusion; I won't have to go to the hospital to receive my doses.

But these drugs are not FDA-approved for my rare mutation so we need to convince my insurance company to pay for them. This may be a serious challenge because the scientific evidence that they will work for me is scant, at best. And the treatment is going to cost a fortune: hundreds of thousands of dollars. Dr. Atkins predicts that the insurance company will deny his first request, and within a few days, it does just that. Jake's mother and her husband offer to cover the full cost of the drugs, and Mirek's mother, in Poland, wants to send us her life savings. But Dr. Atkins suggests we wait. He's hopeful that he'll find scientific support to get me the drugs for free or at a minimal cost.

Dr. Atkins writes a detailed letter explaining that my rare *BRAF* mutation warrants treatment with these drugs. We wait a day, then two. Then another day. On the fourth or fifth day, Dr. Atkins calls me: The drug company has agreed to give me the drugs for "compassionate use." This term refers to using a new, unapproved drug to treat a patient when there are no other options. In other words, *She's going to die anyway, and there's a slight chance this might help, so why don't we try it as a last resort?* The treatment will be free.

Within a few days, I receive two boxes, one the size of a countertop refrigerator filled with ice and my very pricey dream drug trametinib, and a smaller one containing the dabrafenib. Excited, I take pictures of the boxes. What a joy! Christmas in July!

They have to work—they're far too expensive to fail me.

I immediately swallow the first dose of the pills. And wait.

• • •

A few days go by without much of an effect. Then the rash appears.

Skin inflammation is one of the most common adverse side effects of the trametinib/dabrafenib treatment. This reaction to the drugs is experienced by more than half of patients who take them. The two-drug combination, in contrast to each drug by itself, increases the toxicity. There's only one pleasant side effect, which comes out of the blue: my eyelashes grow very long, lush, and coal black, the bottom lashes brushing the tops of my cheeks.

Since I have insomnia from the steroids, I get maybe two to three hours of sleep at night. I'm very tired and catnap frequently; I add sedatives and sleeping pills to my growing medical arsenal. Yet I continue power-walking every day, as far as eight miles, in the early morning or at sunset to avoid sun and heat. I cannot swim because of the rash and my very dry skin, but I bike from time to time in the early morning, sometimes for an hour and a half. Like a soldier always ready for battle in this protracted war with cancer, I am determined to stay in shape.

By mid-July the rash explodes with a force we had not expected. Scary reddish welts cover large areas of my body, and my skin feels as if it's on fire. Dr. Atkins decreases the dose of dabrafenib by half (since it is likely to be dabrafenib, rather than trametinib, that is causing the rash). A few days later, less than two weeks after I started taking one of the drugs upon which all my hopes are pinned, he directs me to stop taking it entirely because my whole body is covered in horrific splotches. This type of out-of-control rash can actually endanger my life, he tells me.

Still, my mind seems to be working well. I am able to read, keep notes, and hold teleconferences with my colleagues at work.

I am coming back to life. But my family and I don't talk much about what we experienced during my mental decline and crash. We're terrified that it will return without warning.

I'm scheduled for another brain scan on July 21. It will be the first one I've had since the catastrophic MRI on June 19 that revealed the new tumors and swelling in my brain. Strangely, I'm not worried about this upcoming scan. I'm resigned to hearing bad news again, so I continue planning for death. I clean out my closets and drawers, the accumulated stuff of my life. But deep inside, against all odds, I am hoping for a miracle.

On July 21, a few hours after the MRI, Mirek, Kasia, and I gather in a room at the Lombardi Comprehensive Cancer Center and wait for Dr. Atkins to deliver my sentence. The wait is long. It's late afternoon, and we're all very tired. Our anxiety is so intense that we don't talk to one another but stare into the distance, biting our nails, breathing deeply, and sighing.

Finally, Dr. Atkins walks in. He is beaming.

"Great news!" he announces. "It worked!"

Before we can absorb his words, he continues. "All of your tumors shrank considerably or disappeared altogether, and there are no new lesions in your brain," he says. "The trametinib/dabrafenib combo therapy was a success!"

Instead of focusing on this remarkable news, I start to argue.

"Dr. Atkins, how do we know?" I say. "How can we attribute my improvement to dabrafenib and trametinib? I was taking them for such a short time. Is it possible they could work so fast? Maybe what worked was the combination of immunotherapy and radiation—or those two plus the targeted therapy. Oh no, we've lost the chance to really know for sure! We'll never know what the magic bullet was!"

Dr. Atkins gives me a dismissive half-smile. "I don't care what worked and neither should you," he says. "The tumors are disappearing. We should be grateful for that."

I *am* grateful. But the scientist in me is annoyed. Perhaps only another scientist can really understand, but I'm dissatisfied by not getting a precise answer to what succeeded in this unique experiment —the experiment of me.

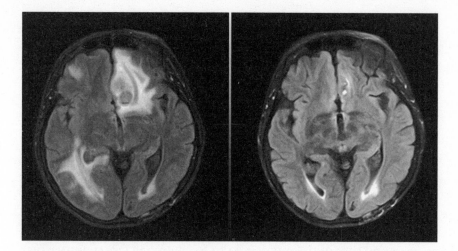

My brain scans from June 19 (left) and July 21 (right). The swelling (the areas in white) has decreased dramatically, and the tumors, including the one in my frontal cortex, have all but disappeared.

Dr. Atkins offers to show us the brain scan on his computer, and Kasia peers at it, amazed at the change.

"It's dramatic," she exclaims. "The tumors have almost all disappeared."

I don't look at the images. I cringe at the thought of seeing photos of my wounded brain. Mirek and I sit in silence, too traumatized to express our happiness. This date marks an amazing breakthrough that we are not ready to trust.

The next morning, July 22, Mirek writes a short note in his diary: *We enjoyed the news as much as we could.*

His diary entry sounds like an unimportant footnote. In truth, we are all in a state of shock. Our emotions have been battered for months. Everyone expected me to die, then I got a pardon, then more bad news, and now another reprieve: the tumors have disappeared.

None of us remembers anything else from that day.

• • •

Dr. Atkins believes it was the dabrafenib/trametinib combination that worked so spectacularly, so he directs me to resume taking a half dose of dabrafenib. The next days and weeks are troubling, as new side effects appear: bleeding sores on my hands, lips, and face. When I awaken at night and go to the bathroom, the image in the mirror is terrifying: blood seeping from my lips has dried around my mouth and onto my neck. I look like a vampire after a busy night. My pillow is stained with blood, and so are the sheets. The skin on my feet is dry and so cracked that every step is painful, and my heels are bleeding too.

On some nights, I have a fever as high as 103 degrees, with chills so profound that in the middle of the hot summer I sleep under two thick quilts and a pile of blankets with a gray wool cap on my head. The shivers nearly knock me off the bed.

There is worse to come. Very early one morning, as Mirek is exercising in the basement, he hears an unusual bang and races upstairs. He finds me unconscious on the bathroom floor, my body slick with sweat, my pajamas drenched. The top of my head is bleeding, and a chair lies toppled next to me. I have fainted and hit my head on the tile wall or stone floor—he can't tell which. I soon come to with no idea what happened. Mirek insists that from that point on, we leave every door inside the house open so he can hear if I'm in distress.

Dr. Atkins decides I need to take a break from dabrafenib again, and then also from trametinib, and he takes me off them. My skin improves and I start feeling well. And, despite my not having any treatment for two weeks, the next MRI, on September 1, shows no new tumors. And the old ones have shrunk even more or disappeared. Every six weeks, I get new brain scans. Over the next few months, several small tumors appear and are treated with CyberKnife radiation. They grow slightly and then shrink. Dr. Atkins keeps me off the dabrafenib but puts me back on the trametinib.

Over the fall of 2015, I continue to have rashes and experience bleeding on my hands, arms, and the top of my skull. But I seem to have returned to the person I was before all of this began. I don't get lost walking around the neighborhood. I remember how to make my favorite recipes. I'm not snapping nonstop at my family. I talk by phone each day with Kasia and Maria in my normal, loving fashion. Mirek and I have friends over for lighthearted dinners. My grandsons come to visit and I play with them happily.

Over time, Mirek shares, bit by bit, how I behaved in June and July. He reveals how I was someone different than the person he knew, a shadow self, and how they all worried the real me was gone forever.

I promise him that I will never again be so mean to him and our family. But inside, I know it's a promise I may not be able to keep if my brain fails me once more.

From time to time, I make stupid jokes, pretending to lose my mind, faking that I don't know where I am. He doesn't laugh. It's cruel, I realize, and I stop doing it. After all, I'm the only one who didn't witness what happened. I'm the one—in a way—who suffered least.

In January 2016, after a year of aggressive cancer treatments, the torment of new scans, and anxiety over whether they would reveal new tumors in my brain, I sit on the couch in my living room. My arm is swollen and tender. It's the lymphedema from my breast cancer, exacerbated by immunotherapy for melanoma.

Why haven't I done something about this sooner? I can't believe I've never taken care of it.

I do a computer search for nearby clinics with physical therapists who specialize in lymphedema. *Oh, here's one close by, at Inova Fairfax Hospital.* I pick up the phone and call the office. The receptionist schedules me for an appointment in just a few days, on January 15. I wait patiently for the date to come.

The morning of January 15, I use Waze to guide me to Inova Fairfax Hospital and proceed to a parking garage. There are no empty spots so I climb to the highest level and park on the open floor. I get out of the car and look around.

This feels so familiar . . .

I have a strange sense that I've been here before. But when it was, I just cannot remember.

I take the stairs down to the first floor and follow signs to the hospital, which seem complicated: up and down, left and right. These corridors, these elevators, these signs . . .

Have I been here before?

With each step I take, a feeling of uneasiness and mystery intensifies. I reach the receptionist's desk in the waiting room. From my fogged memory, I seem to recall these places but cannot remember the circumstances that might have brought me here. After a while, I hear my name called. I lift my head and see a woman in the doorway.

"Oh my gosh, it's you!" she exclaims. "I was *sure* you would never come back."

I vaguely recognize her. Slowly, as if from some previous life, I recall her name: Theresa. When we walk into the therapy room, I recognize it, too, in a fuzzy, imprecise way.

Theresa asks me how I've been and what brought me back.

I try to explain. I tell her about my illness and treatments, about the tumors in my brain. I tell her that until just this moment, I didn't have the slightest shred of memory that I'd visited this hospital before. I tell her I recognize her face and recall her name but not much else.

She smiles.

"We were all very sure you would never return," Theresa says. "At your last visit, you were so angry and so dismissive of our advice. I told the staff we'd never see you again."

When I cringe, she quickly adds, "I'm so happy you came back."

Now the memories rush in. I recall arguing with her, my rude tirade, my refusal to listen. I remember storming out in anger.

I apologize over and over. I feel terrible about how I acted, but she comforts me.

"I understand," she says kindly. "I've had patients before who refuse to go through this treatment because it doesn't feel or look right to them. They prefer to suffer." She takes a look at my arm. "Let's get back to work."

I sign up for twelve PT sessions, and over the next two months, I follow her instructions with devotion. I learn how to bandage my arm and order the special lymphedema garments that I need. I do everything she tells me, and my arm improves dramatically. One day, with a sly grin, Theresa tells me I'm her "most improved patient."

As we work together for my recovery, Theresa and her colleagues become my dear friends. When I finally finish my sessions, we have tears in our eyes as we hug goodbye.

I am now remembering other things that happened in that period, albeit hazily. I recall the young pest-control guy who visited us—and how I angrily fired him when he couldn't tell me what was in the chemical sprays he used. I remember the day I was lost in the street and peed all over myself.

And chanterelles are forever ruined for me. Once they were my favorite mushroom and one of my favorite foods, an emotional connection to Poland and a special part of my childhood; now, I can barely say the word aloud. The very name transports me back to that terrible day in the park. They prompt a kind of traumatic response, not just for me but for my family. Later, once I realized that my behavior that day was part of my mental breakdown, I begin to associate chanterelles with losing my mind. I worry it will happen again. It's a fear that haunts me every day.

Nearly a year after the incident, when he can finally bear to talk about it, Mirek tells me that on that morning, he and Kasia worried that it wasn't safe for me to walk briskly for seven-plus miles in the park. But when I insisted I was fine, they had ample reason to believe me, he says. Six weeks after the surgery in January 2015 to remove the brain tumor that had affected my vision, I had radiation to kill any lingering cancer cells around the surgical site and two tumors. The very next day, Mirek and I were on a twelve-hour flight from Washington, DC, to Hawaii, where we cycled two hundred miles and I participated in a 5K race.

Before we made the long trip, we'd asked Dr. Aizer, my radiation oncologist at the Brigham, if I'd be okay. "Absolutely! Enjoy yourself!" he'd answered. He was right; I suffered no ill effects from our ambitious vacation. A few weeks after that, I went cross-country skiing in New England with no problem. It was typical of me. In 2010, in the midst of chemotherapy for breast cancer, I'd gone downhill skiing in Colorado at an elevation of fourteen thousand feet, a helmet covering my bald head, and one arm so swollen from lymphedema that I was barely able to grip a ski pole.

With that personal history, it never occurred to me to rest after the CyberKnife procedure. And my family believed me when I said I was strong enough. To all of us, it seemed clear: walking in the park would be, well, a walk in the park.

Kasia now says that they all were so desperate to believe I would be okay—that I wasn't going to die—that even she, a medical doctor, chose to ignore her concerns. "We wanted so much to restore the normal order of things, to get back to our normal life," she says.

Of course, for other families not as obsessed with fitness as ours, the decision to exercise vigorously in the woods that day probably seems crazy. But it wasn't crazy for us. My insistence that we work out was completely in line with my driven personality and my role in the family as the one who set the schedule. I didn't become someone entirely new. In fact, the opposite happened: I insisted on being myself despite the cancer and radiation.

My children and Cheyenne tell me now that they regret letting me drive the car to find Mirek amid the chanterelles. They should have insisted one of them take over, they say. But I was so angry at that moment that they were afraid they'd escalate the tension by objecting. "I figured as long as you weren't going on the highway, that you were on a park road practically empty of cars, it would be okay," Witek says.

What he mostly remembers, he says, is how much he worried that my unpleasant behavior would be the new normal for his mother. Worst of all, he thought that that unloving version of me would be the last one they would live with before I died.

Our family, like so many others affected by mental illness, struggled to adjust to the new normal. As my husband and children found when I became mentally impaired, making that adjustment was extremely difficult. It was hard for them to recognize my personality changes, especially because I insisted I was fine. Even as the changes became more obvious, my family remained in denial because the new normal was so disturbing. It pained them that their mother or wife couldn't function the way they were all used to. Accepting that things were different meant that my family would have to alter its long-standing way of operating and that someone else would have to step into my role as the one always in charge. If I could no longer function in that role, moreover, who was going to tell me? How would they take that responsibility away from me? Who would assume my position in the family structure, and how much would I resist? Could they force me?

In my family, no one wanted our happy life to change. So we all refused to accept the full reality of my illness. Training for triathlons! Chanterelle hunting! These are the things that we love, and so off we went to the park that day as if I'd not just learned that I soon might die. I suppose you could argue that the exercise helped relieve stress, and it certainly did. But that's not the main reason we headed out for the day. We did it because it's what we always did—and we did not want to accept that anything was different.

If a loved one or coworker suddenly slumps over and is paralyzed on one side of his body, most of us would recognize symptoms of stroke and immediately call 911. Acute symptoms like this are easy to see. But behavioral changes can be much harder to recognize and accept as alarming or serious. This is especially true if they come on slowly, like gradual memory loss or small changes in someone's physical abilities. We tell ourselves, "Mom is just getting older—of course she's forgetting," or "Her joints ache, that's the reason she's no longer a sweet and loving person." It can be very difficult to admit that personality distortions like the ones I experienced—anger and irritability, loss of inhibition, and lack of empathy—might be signs of serious physical problems in the brain and that a physician's help is needed.

When I became very angry in the park, my family could tell something was off but they also felt there was little they could do about it. I was tired and grumpy, an exaggerated version of my usual type A self —but nothing so extreme as to set off serious alarms. They did ask me to take it easy. But was I listening? That night, I was the one who cooked dinner for the family even when it became clear I was struggling and couldn't navigate my way around my own kitchen—because that was my role, and I had no intention of giving it up.

11

Survivor

Despite all my years of studying brain disorders, for the first time in my life I realize how profoundly unsettling it is to have a mind that does not function. And the more I remember from the days and weeks of my madness, the more frightened I become that I will lose my mind again. Perhaps *madness* is not the proper term to describe my condition at the time. After all, it is not an official diagnosis, but it is often used informally to mean mental instability, insanity, and angry and disorganized behavior. So instead, I think of myself as having experienced a number of symptoms connected to a range of mental disorders. In other words, I had a brush with insanity.

And I have come back.

Despite conducting research on mental illness for over thirty years, I believe it is my own suffering that truly taught me how the brain works—and how profoundly frightening it is when our minds fail. I personally experienced how scary it is to live in a world that makes no sense, where there is no logic because the past is quickly forgotten and the future can't be planned or foreseen. As a result, I have become preoccupied with examining my own mind. I continuously test myself to see if I'm slipping again. I solve math problems, try to remember dates, check to see if I've forgotten any loose ends or details. I exercise my own mind like I'm training for a marathon; I try to be more curious, inquisitive, sharp, and logical in an effort to make up for any losses that I may have experienced. I do this because I live in constant fear that my insanity will return.

And—to memorialize my experience—I write, write, write. I feel the overwhelming urge to share my experiences with other people. By sharing, I relieve my own fears and perhaps soothe those of others. These are my new obsessions.

On March 13, 2016, a little over a year after I was first diagnosed with metastatic melanoma, the *Sunday New York Times* publishes my essay "The Neuroscientist Who Lost Her Mind." The response is immediate and overwhelming. I receive over two hundred e-mails from people all over the world, thanking me for writing so honestly about my experience with mental illness, and my piece is one of the most e-mailed articles in the *New York Times* that week. Many people with mental illness and their family members write to me. Physicians who work in the field thank me for shining a light on this issue. Dr. Thomas R. Insel, a former director of the NIMH, writes to me, "You have done something so important for people with serious mental illness who do not have an observable lesion. Not only have you reminded us all that mental illnesses are brain illnesses, you have reminded us to be hopeful. People recover."

What is it about this article that struck such a chord with so many people?

Brains fascinate us with their complexity and the mysteries they hold. Everything we dream, think, feel, and do—everything that makes us who we are—comes from the brain. We *are* our brains. It is terrifying when the mind breaks down due to illness or aging, and we lose what is dearest to all of us and to our loved ones: our personas. People are hungry to know more about the mind and the mental ailments that we all hope can one day be explained and cured.

In April 2016, an ordinary envelope arrives in the mail. When I open it, I'm astonished to find I have a remarkable, once-unimaginable new title for myself: cancer survivor. On May 6, 2016, the Lombardi Comprehensive Cancer Center is holding its annual melanoma survivorship luncheon, and I am among the guests invited by Dr. Atkins and his team.

Survivor. Am I a survivor? They must have made a mistake. I have not been cured. At best, I am in remission. True, I'm still alive sixteen months after my diagnosis, which is an incredible feat given the grim prognosis of my disease: four to seven months. But I'm still suffering from rashes all over my skin. And who knows how many cancer cells continue to lurk undetected in my body, waiting to sprout into tumors?

But there it is, in an official letter, an honor more precious—and unexpected—than any I can recall.

What does *survivor* mean? What does it take to be admitted into this special club?

In the days leading up to the event, I ask myself many times about my surprising new identity. I'm curious about what it really signifies. At its most basic, a survivor is someone who was seriously ill but isn't dead, at least for now. Not a bad label, given the alternative, but somehow less than entirely satisfactory. Or maybe *survivor* includes everyone currently free of any detectable signs of the disease. For me, this definition seems too random, too dependent on the accuracy of our current diagnostic tools. Melanoma cells can lay dormant and hidden for years, biding their time and then emerging from the depths of the body when the conditions are right to strike and kill quickly. *Survivor* is a problematic category if it simply means that someone's cancer cannot be detected with the tools available at the point in time when the hosts mailed out the luncheon invitations.

I Google the general definition of the word and learn that *survivor* means a person who remains alive, carries on despite hardships or trauma, perseveres, and remains functional or usable. This sounds a lot more inspiring, especially the last part, "functional and usable."

Am I functional and usable? What about the others who attend? How incapacitated will they be? Are they still functional and usable?

I start obsessing over this idea and begin examining my own life, all that I have done and been, the good and the bad. I think about the people I love, especially the two people I brought into this world and nurtured, Kasia and Witek. Am I a successful human being? What have I achieved? Do I measure my life by my career success, the hundreds of scientific lectures I've given and articles I've had published? Or is my real achievement my devotion to my family, who in turn have stood by me through the wintry days of gloom and tragedy? I reflect on my cheerful, still so innocent grandsons, Sebastian and Lucian, who always wait on the front porch for their loving *babcia* to arrive from Washington, DC.

But I've also failed. I still carry guilt and regret for the breakup of my first marriage and for not being there to support my first husband through his unsuccessful battle with melanoma. And what about who I am now? Am I functional? Usable?

The day of the luncheon is dreary, rainy, and cold. I'm not sure I want to go there and mingle with people I don't know, people who were—or maybe are—on the verge of dying. But I shake off my reluctance, and off we trek, Mirek, Witek, Cheyenne, and I.

More than seventy people crowd into a conference room at Georgetown University Hospital: Dr. Atkins and other doctors and nurses and about thirty melanoma patients along with their families and friends. I recognize many of these faces from my visits to the cancer center, although in most cases, at the time I had no idea that they also were suffering from melanoma. Today, everyone looks healthy, and we're all smiling.

The survivors range in age from their late thirties to over eighty, with most in their sixties, I'm guessing. And almost all of them are eager to share their stories: symptoms, diagnosis, treatment. Like soldiers emerging alive from a battlefield with emotions still raw, they easily discuss their experiences with comrades who've undergone similar hardships, the only ones who really understand.

One woman tells us she was diagnosed fifteen years ago with early-stage melanoma. Unfortunately, in recent years the disease metastasized throughout her body including her spine. Immunotherapy saved her life but she has trouble walking. She is wild type, meaning she doesn't have a mutation in *BRAF*, the melanoma-related gene, like I do, so the targeted treatment I got won't work for her. She tells her story with a smile as her husband holds her hand.

A tall man of about seventy, a retired doctor, was diagnosed with advanced melanoma over six years ago. It didn't appear first on his skin—which is unusual but not unheard of—but instead attacked from inside his body. He smiles as he describes being saved by the Georgetown team and how well he's feeling now. A stout and healthy-looking gentleman of about the same age brags about the number of beers he drinks on a weekday (over twenty) and on the weekends (over thirty) and tells us about his beloved horses and chickens on the Southern farm where he lives. He has undergone various harsh treatments for advanced melanoma, some without much success, but the latest immunotherapy has worked for him, although he's developed another kind of cancer. Unfazed by these adversities, he says he looks forward to horseback riding and drinking. A couple seated at the far end of our table has come all the way from Florida, where they'd retired just weeks before the wife's melanoma diagnosis. The Florida doctors told them she would die soon because there were no viable treatment options. But she found the immunotherapy trial at Georgetown, which has been successful so far, and now they commute to the Lombardi center every few months so she can get check-ups and scans before returning to golf in the Florida sunshine.

We watch two short videos of other successful cases—other survivors. A woman in her forties describes how she discovered a large tumor on her thigh that turned out to be melanoma and was informed by her doctor that she would soon die. As she tells her story, her two very young daughters and a stepson giggle and play and hug her. She limps slightly, smiling shyly. A man well over eighty developed a large, scary-looking tumor on the scalp of his bald head. After immunotherapy, he says, his tumor disappeared as if a magic wand had touched his scalp.

As we mingle with the guests, I recognize Dr. Atkins's nurse Bridget, whom I first met when I entered the clinical trial a year ago. She compliments me on how healthy I look.

"Do you remember that day in Dr. Atkins's office when you all gathered around me and broke the terrible news that my brain tumors had grown and were pressing on my brain?" I ask her. "That there seemed to be no hope? And then you started to cry?"

"I will never forget it," she says. "I'm so sorry I cried. I should have stepped out of the office."

"No, no," I say. "It was so human, and strangely, it gave me strength to see that other people do care, do feel for me, and would be sad if I died. We are social animals. We should feel for each other, cry for each other. There's nothing wrong with showing our emotions. I only wish that it happened more often."

I speak briefly with the wife of a survivor. Her husband, a grandfather of eight-month-old twins, had tumors that quickly disappeared after immunotherapy. She tells me that she's so happy that he will have the chance to get to know the twins and enjoy being a grandfather. "He is such an optimist," she says. "I've seen him suffering badly from the side effects of the drugs. He almost died from them but he never complained."

Dr. Atkins gives a short presentation about the immunotherapy with which we, the survivors, have been treated. The immunotherapy clinical trial is very successful, he says; the vast majority of us survivors are expected to live for some time. And only one patient in the trial has died, he adds.

"Several years ago there would be no luncheon like this," Dr. Atkins says, "because most of you would likely be dead." His words may strike some people in the room as harsh but he is telling the truth: were it not for the immunotherapy he administered, I would certainly not be here today, and the same goes for many of the people gathered here. Before this miraculous new treatment, the majority of patients with advanced melanoma had no chance of survival. Immunotherapy is, indeed, a miracle cure, and not just for melanoma but for a number of other cancers too. It doesn't work for everybody yet, and it may not work forever except for the luckiest of patients. But it works. We survivors of advanced melanoma are living testimony.

We have many questions when he finishes; mostly, of course, about our own fates. How can we make sure that the disease does not come back? "There are no guarantees. You'll need to come frequently for medical checkups," he says. Since there's a hereditary link in melanoma, what can we do to protect our children? "Currently there is nothing we can do but protect our children from sun and make sure they always wear sunscreen," he advises. Do a positive attitude and a strong will to live affect survival? "Perhaps," he says. "They certainly don't hurt. We don't know very much about the influence of will on survival." How can other melanoma patients—those not lucky enough to be in the clinical trial—afford the extraordinarily expensive immunotherapy drugs? "We have no answers to that yet," he says. "It obviously depends on the insurance you carry." How can patients deal with toxic, sometimes life-threatening side effects of this treatment? "We are trying to provide as much expertise from other medical fields as possible to deal with side effects, but sometimes this help is inadequate," he says.

A photographer takes pictures of all of us with Dr. Atkins and his medical team. It feels like a graduation photo. We have persevered. We have remained functional or usable. We are true survivors.

At the end of May 2016, after several scans that reveal no new tumors, I stop taking trametinib. It's a huge relief—and a big cause for concern. The horrible rash that I've continued to suffer from disappears almost instantly, and I feel better. But what will happen inside my skull now that I'm off medication? Will the tumors resurge and attack? Dr. Atkins seems confident that the melanoma cells in my whole body have been defeated and that they have, as he says, "stopped seeding"—spreading through my bloodstream into the rest of my body. It's reassuring to hear that my cancer may be eradicated for good. But without the drugs I feel like a white-water kayaker without a life jacket.

Another tumor appears at the end of July 2016, after I've gone several months without treatment. It's in my cerebellum, the region that controls voluntary motor movements, but it's so small it causes no symptoms. A few weeks later it gets zapped with the CyberKnife.

Throughout the summer of 2016, I slowly return to my former self. I run and swim and cycle, and I travel with Mirek to visit my family in their own homes. It's a nice change to be able to go away, so heartening to no longer be the seriously ill mother and sister they came to see as often as possible—never knowing whether each time would be the last.

But although I am tumor-free for now, another disaster is brewing in my head: brain tissue necrosis, the delayed and potentially fatal effect of radiation. Necrosis occurs when an area of dead tissue forms at the site of a brain tumor after radiation therapy and the surrounding tissue doesn't heal. It is more prevalent among cancer patients these days than it once was because of the increasing use of SRS and CyberKnife techniques combined with immunotherapy; the two procedures work synergistically to destroy tumors but also destroy surrounding healthy tissues.

The symptoms of brain tissue necrosis may not appear until a year or longer after radiation treatments. It's been fourteen months since I received a series of radiations for my tumors. So in a sense I am right on time when, at the end of August 2016, the site of the biggest tumor, in my frontal cortex, begins to act up.

As I'm getting ready for a hiking trip in the White Mountains in New Hampshire with Maria, I notice a blind spot at the top of the visual field in my left eye. At first, I don't pay it much attention. Maybe it's a slight cataract, I think, and I try to ignore it into oblivion. But within a few days the vision in my left eye deteriorates as rapidly as if a curtain is being lowered from the top to the bottom of my eye. From day to day, it gets worse and worse. My doctor orders an emergency MRI of my brain and eyeballs. The scans confirm what we suspected: the problem is not the eye itself but my optic nerve. Aftereffects of radiation to the frontal cortical tumor, which was very close to my left optic nerve, have destroyed the nerve. I am diagnosed with irreversible optic neuropathy, total blindness in my left eye. There is no cure. I will have to learn to live with one eye.

Two days later, I fly to Boston and meet my sister. We are ready for a three-day hike. At the last moment, I decide to buy trekking poles at REI in case I have trouble with my balance. They are very light and comfortable and turn out to be my lifesavers during our challenging trip. We hike up rocky and steep Mount Washington. Being blind in my left eye, I no longer have depth perception. At the start of our trek, it's very hard for me to estimate the inclines, and I fall frequently. I have trouble climbing, and it's even worse going downhill. I trip and stumble. But in very short order, I adjust. We successfully and joyously complete our planned route through the mountains and our three days of hiking.

Back home in Virginia, I have to relearn so much. How to run without stumbling—many times I return from my daily run with bloody knees and palms. How to cycle—I add a side mirror to my bike so I won't crash into objects on my left. How to type and read in my new off-center world. How to drive my car—I rotate my head so far around before changing lanes that Mirek jokes I'm becoming an owl. I also learn to ski without depth perception. I downgrade myself from expert-level, double-diamond runs to single-diamond trails. Fortunately, swimming is easy. I don't bump into anything but water and just follow the line at the bottom of the pool.

It's a slow process, but my memory continues to come back, especially as I begin to write this book in the spring of 2016. As I reflect on the two months of my journey and try to piece together what happened, I can grasp little bits here and there and often recall entire episodes.

But when I ask my family questions to help me fill in the gaps, they usually don't want to talk about it. Mostly they say they can't recall, and I think they are telling the truth. It is too traumatic for them to relive. It's as if they don't want to resurrect the version of me who was so unkind, the version they worried would be their final memory of me.

In the spring of 2017, Kasia asks Sebastian if he remembers the time I was mean to him. It's been two years since it happened, and Sebastian is now ten years old. He's grown into a tall, thin kid with a tremendous talent for running. He tells his mom he doesn't know what she's talking about. He doesn't remember anything like that at all.

In truth, conjuring up the images of those episodes isn't easy for me either. I'm still embarrassed at how I treated Theresa during my first physical therapy appointment, even though I couldn't help it and she immediately forgave me. I wince at how I behaved with Sebastian and Kasia and Witek. And with Mirek, especially. There is trauma still embedded in my mind, a fear that I may again explode without any warning and become a brute whom people will want to avoid. These worries that I won't be able to control my behaviors, that there is so much unpredictability lurking in me, don't go away. They're now part of who I am.

These days, long after the Nina Simone documentary and the grand opening of the grocery store, I still tremble at the memories of those lights and noises, the loud music, the sharp whites of piercing life, the black ghosts of death. During that emotional movie, a death thought came upon me like a hungry tiger. Throughout my ordeal, I was never consciously afraid of dying, and I believed that death was simply our longest sleep, absent of nightmares. No pleasures, nothing. Yet looking back, I'm surprised that I remained calm and composed through so many near-death experiences. The fact that I was so often not fully aware of what was happening to me was a kind of protective obliviousness, I'm sure. But during the infrequent periods when I reflected on the thought that I might die soon, I knew that I had lived a fulfilling life, and that insight gave me strength and peace. Today, as before, my passion to stay alive and my readiness to die are intermingled.

I continue to worry about my mind. My brain will never be as it was before. It has been injured by tumors, shot through with radiation, assaulted with drugs. It is scarred, figuratively and literally. And since my brain is different, I am not exactly the same person that I was before the illness. But strangely, I feel completely myself. Perhaps my brain has resculpted its damaged connections or rerouted them in some grand effort to reinstate their original structure and function. Or maybe I just don't see the changes in myself, in this new normal, in the new self that I have embraced. My family thinks that the truth probably lies somewhere in between—but we will never know.

In one regard, at least, I am different than I was before: I've become more aware of living. I try harder than ever to find meaning in ordinary things every day. When I look at trees swaying in the wind, petals from the blooming bushes in our yard scattered on the ground, I think, *The world is so beautiful. I'm so happy I'm living when I could be dead.*

In the foreseeable future, and perhaps for as long as I live, there will be more brain scans, more tests, and the anxiety of waiting for results. There may be unexpected, undesirable findings followed by more treatments. I am competing with a particularly wicked and perverse opponent, an illness that is very hard to beat. It feels like an Ironman competition that demands, in addition to the latest scientific achievements, an iron will, body, and mind. In this race, I'm not rushing to the finish line because there isn't one. There are no medals or trophies to earn, no accolades, no cheering. There is only the deep satisfaction of another day lived, another day with the people I love.

Epilogue

I resolved not to compete in any races, at least not in the near future, in order to focus on healing, family, and work. But in December of 2016, our family decides we will register for the Quassy Revolution3 Triathlon in Middlebury, Connecticut, a competition known as "the Beast of the Northeast." Held each June, it's a particularly tough race: 70.3 miles total, as long as a half Ironman, with hilly cycling, running, and a 1.2-mile swim in a frigid lake. We've never before attempted anything this difficult.

At first, I'm reluctant to begin planning. I wonder if I'm kidding myself about being able to compete in an athletic competition. What if, in the next few months, I develop new tumors? What if my brain swells again? How can I be sure that I'll be in good enough physical shape by June to take this on—or that I'll even be alive? But I don't share my fears with anyone. The rest of the family is so excited about the prospect of all of us—me, in particular—returning to the races that after a few days, I give in and start to train.

I know that I won't be able to do the entire triathlon by myself, as I planned to do back in January 2015, before my brain tumors were discovered. I just don't have the strength or stamina. So we decide that three of us will compete as a team, each of us taking one of the events: Mirek will cycle, Jake will run, and I will swim. My grandsons, Lucian and Sebastian, are excited to race in the kids' triathlon, and Kasia will compete in the entire Quassy half Ironman by herself.

I spend the winter of 2016 in training. I swim four times a week at a nearby pool, and several days a week I also cycle on an indoor bike and run in order to boost my energy, increase my overall strength, and attempt to return to the form I had before disaster struck. Getting back into shape is much harder than I imagined. Even though I stayed active throughout my illness, taking a long walk nearly every day and often running too, my muscles are weak. Physically, I am not the same person I used to be. I don't have the same flexibility and balance, and with sight in only one eye, my vision is poor. Since I can't see well, I get easily disoriented, not only in new environments but even on the familiar trails behind my house, where the ground is uneven and creeping vines trip me up.

Despite my own doubts, over the weeks and months I continue my daily training sessions. I love to tie the shoelaces of my sneakers and run out into the chilly morning when the sun is just starting to peek through the trees and the birds are screaming their crazy tunes. When spring arrives, the intoxicating fragrance of lilacs almost knocks me out the moment I open my front door. Each day, I increase my distance and speed. I come back from these morning runs aching and tired but beaming with joy, and I gulp my hot coffee with an almond croissant, my reward.

At the pool, I love to slip on my goggles, dive into the deep water, and swim. My arms slice through silky water, my lungs open wide to draw in the air, my rhythmic, powerful strokes propel me forward. Day by day it gets easier, smoother, becoming almost effortless. I'm not nearly as fast as I used to be, but the pleasure of the water caressing my body and the feeling of accomplishment are just the same as before.

And then, suddenly, from nowhere, trouble strikes again.

One afternoon in May 2017, two weeks before the race, I'm sitting in my office at NIMH when my left leg starts twitching uncontrollably. I try to hold it still but can't. Although it's a very short episode, lasting maybe thirty seconds, I am very frightened. I know what it means: I've had a small seizure. I immediately go in for an MRI, which reveals a small but disturbing crater in part of my right motor cortex, the site that controls the motor movements of my left leg and arm. Radiated almost two years ago, the spot has now turned into necrotic tissue, with dead cells and debris that are choking off healthy brain cells. That's why my leg began twitching.

Necrosis is a side effect of radiation, and it's not good news. My brain is not healing well. My first reaction is that I will have to cancel the race in order to focus on healing my brain.

Dr. Atkins prescribes steroids yet again for the inflammation and swelling in my brain that have occurred as a result of the necrosis. And he explains his long-term plan for healing the wounded brain tissue. Every three weeks, I will receive IV infusions of a drug called Avastin, which was originally developed to treat solid cancer tumors by choking off their blood supply so they stop growing. I don't have any new tumors, but Avastin will, Dr. Atkins hopes, seal the leaking blood vessels in my brain and stop the edema and inflammation in the wounded tissue. Nobody really knows whether it will work, he adds; Avastin has been used only occasionally to heal post-radiation wounds like mine, and the outcomes are not yet clear. But there is no other treatment to try, he tells us, so we must simply hope for the best.

When I mention the upcoming Quassy triathlon, Dr. Atkins says he doesn't want me to swim in the lake. He asks a rhetorical question: "What if you have a seizure in the water?"

I weigh my options, and after a few days, I decide I will not cancel. I am going to swim my 1.2 miles. I call the organizers of the race and ask for help in securing a guide who can swim next to me in the lake and make sure I'm safe. A volunteer involved in the logistics of the triathlon, Daniel DeHoyos, calls me and offers to swim with me. "It would be my honor," he says. "I've read your essay in the *New York Times*. What an extraordinary journey you've gone through." Witek also rushes in to help, offering to join me the day before the competition for the training swim that competitors take to scout out the route.

The race is scheduled for Sunday, June 4—Kasia's birthday—and bad weather is in the forecast. On Saturday, June 3, Mirek and I make our way north by car from Virginia to Connecticut. Gray clouds roll in and a light drizzle falls. It's becoming colder and colder. That afternoon, we reach Waterbury, Connecticut, and check into the Hampton Inn. Both Mirek and I are anxious about the potential hazards that tomorrow may bring: hilly roads slippery with rain; cold lake water that could trigger my seizures; the long distances with significant physical challenges that we each will have to endure. But we continue forward on our path of no return, and we try our strength that afternoon in training runs along the triathlon route. We drive to nearby Quassy Amusement Park. Kasia meets us there, and she and Mirek quickly disappear into the hills on their bikes. Witek has just arrived from Pittsburgh, and, accompanied by my son, I dip into the water.

I am wearing a long-sleeved wetsuit. The water is not that cold! It's fragrant and sweet. The lake is choppy with light waves but so beautiful, framed by the green forest up to the horizon, where mountains rise. My swim with Witek—we do a couple hundred yards of steady, purposeful strokes—is delightful. When Mirek and Kasia return from their bike ride, they tell us it was a bit scary—the roads are treacherous, with sharp inclines and declines, and wet from recent rains, but at least now they know what to expect tomorrow.

That night, still anxious about our fate, Mirek and I cannot sleep. By 4:30 a.m., with the sounds of other Quassy competitors awakening in the room above us and stirring in the halls outside, we get up and get ready. After a light breakfast, we drive to the lake, arriving shortly after sunrise and getting a good spot at the already crowded parking lot at the lake.

Last night's rain has stopped, and the morning is chilly but calm. The first rays of sun emerge from the clouds and color the lake with a golden hue. The water surface looks like honey; smooth, undisturbed, glistening in the morning light. We gather our gear and head toward our positions. The swim is the first leg of the competition, followed by the bike ride, and then, finally, the run. Mirek pumps up his bike tires one last time in the spot where he will wait for me when I emerge from the water. I once again check the two-hundred-yard route I'll take from the little sandy beach to our transition area, where I will hand him the timing chip that keeps track of each team, and where we'll kiss each other goodbye. I retrace the route several more times to be sure I won't get lost on the sprint from the lake to Mirek.

Among the hundreds of competitors gathered on the beach, I find Daniel DeHoyos waiting for me. He is tall and muscular, and his friendly strength gives me confidence. And here is Kasia in her black wetsuit! We look like a herd of seals on this tiny beach. I stand out, as I'm in a special red cap that is given to people who may experience distress during the swim. Still, I feel so proud to be one of them. I'm scheduled to start my swim in the second-to-last cohort (or wave) of competitors. Kasia will hit the lake five minutes after me, in the final group of swimmers.

When I get ready to dive into the water, I hear an announcement through a loudspeaker: "Barbara Lipska, a multiple cancer survivor, is starting now!" A quick thought passes through my brain: *This must*

Waiting for the start of the Quassy triathlon with Daniel DeHoyos and Kasia.

be Jake's doing, this publicity stunt! Two weeks before the race, Jake wrote an article about our unusual team for the *Wall Street Journal*, "A Triathlon Is Easy Next to Soviets and Polio." It was a beautiful tribute to Mirek and me and our family. (Only after the race do I learn it was Daniel's idea to make this announcement!)

People are cheering for me as I jump into the water! And then I hear only splashing, arms cutting through the water, legs kicking. I try not to lose sight of Daniel, who is swimming right in front of me with a red rescue buoy attached by a rope to his powerful torso. It feels great to follow him so easily, to be safe in his presence.

At the huge orange buoy that marks the first turn for swimmers, Kasia appears in the water right next to me. Even though I started before she did, she is already passing me, and she yells, "Mum, are you okay?"

"Of course I am!" I scream back above the din, and I continue my swim.

As I follow Daniel, I begin to feel great—relaxed, and so happy that I am competing in a real race. It takes me fifty minutes to swim the 1.2 miles. When we reach shallow water, Daniel and I stand and embrace each other as the small crowd at the beach screams and cheers for us again.

I run to Mirek as fast as I can. He kisses me and grabs our timing chip, then hugs and thanks Daniel.

"Life is a team sport!" Mirek says, beaming with joy. As he takes off on his bike, he turns back and shouts to us: "And remember, my love, we'll conquer this beast!"

Acknowledgments

Thank you to my family for always standing by my side and caring for me in the hardest of times, especially my husband, Mirek Gorski. Thank you to my children, Kasia Lipska and Witek Lipski, for your love and always being there for me. Thank you to my sister, Maria Czerminska, for showing amazing dedication to finding the best options to save my life. Thank you to Jake Halpern and Cheyenne Noble, the loving spouses of my children, and to my brother-in-law, Ryszard Czerminski, for your unwavering support. Thank you also, Jake, for your encouragement and help with the op-ed article I wrote for the *New York Times*, without which this book would not exist, and for introducing me to Elaine McArdle, my co-writer and now a dear friend.

Thank you, Agata and Jason Ketterick and Jan Czerminski, for quietly cheering for my survival. Thank you also to my devoted extended family, Tamar Halpern and Paul Zuydhoek, for being such good friends when I needed you most, and Steven Halpern and Betty Stanton, for your kindness and support. And last but not least, thank you to my brilliant grandsons, Lucian and Sebastian, who kept me going in my darkest hours.

I'd like to express my gratitude to the doctors who treated and cared for me: my wonderful family doctor of nearly thirty years, Dr. Eugene Shmorhun; Dr. Michael Atkins at the Georgetown Lombardi Comprehensive Cancer Center in Washington, DC, and his team, in particular Kellie Gardner; the team at Dana-Farber Cancer Institute in Boston, in particular Dr. Stephen Hodi, the director of the Melanoma Center and of the Center for Immuno-Oncology; my neurosurgeon Dr. Ian Dunn at Brigham and Women's Hospital in Boston; and especially the outstanding radiation oncologist Dr. Ayal A. Aizer.

Thank you also to my wonderful physical therapist Theresa Bell.

A special thank-you to my friend Dr. George E. Jaskiw for his review of this book as it developed. Thank you to a number of other physicians who helped with various sections of this book, including Drs. Bradford C. Dickerson, Erica Swegler, Jason Karlawish, Éric Fombonne, and Wendell Pahls. We also appreciate the help of Susan L.-J. Dickinson, executive director of the Association for Frontotemporal Degeneration, and Warren Fried of the Dyspraxia Foundation USA.

I am also very thankful to my colleagues at the Division of Intramural Research Programs of the National Institute of Mental Health for believing in me and my recovery, and to my coworkers and friends at the Human Brain Collection Core, NIMH. Thank you especially to Dr. Susan Amara, the NIMH scientific director, and Dr. Maryland Pao, the NIMH clinical director, as well as to Gwendolyn Shinko, the NIMH administrative director.

My co-author and I want to thank Leora Herrmann for her encouragement, and a very special thank you to Jack McGrail for his unflagging love and support.

We want to thank our agents Esmond Harmsworth and Nan Thornton at Aevitas Creative Management for their guidance, support, and good humor.

Thank you to our wonderful editor, Alex Littlefield, who believed in this project from the start, as well as to Pilar Garcia-Brown and everyone at Houghton Mifflin Harcourt.

Notes

Prologue

page

xiii   *experiences a mental illness:* Z. Steel et al., "The Global Prevalence of Common Mental Disorders: A Systematic Review and Meta-Analysis, 1980–2013," *International Journal of Epidemiology* 43, no. 2 (April 2014): 476–93, https://www.ncbi.nlm.nih.gov/pubmed/24648481.

xiv   *forty-four million adults each year:* National Institute of Mental Health, https://www.nimh.nih.gov/health/statistics/prevalence/any-mental-illness-ami-among-us-adults.shtml.

      27 percent of adults: World Health Organization, http://www.euro.who.int/en/health-topics/noncommunicable-diseases/mental-health/data-and-statistics.

      homeless and incarcerated people suffer from mental illness: https://www.nami.org/Learn-More/Mental-Health-By-the-Numbers.

      $1 trillion each year: https://www.usnews.com/news/best-countries/articles/2016-04-12/who-makes-economic-argument-for-mental-health-treatment.

$193.2 billion in the United States: https://www.nimh.nih.gov/news/sci ence-news/2008/mental-disorders-cost-society-billions-in-unearned-income.shtml.

who die each year by suicide: World Health Organization, http://www .who.int/mental_health/prevention/suicide/suicideprevent/en/.

suffer from mental illness: https://www.nami.org/Learn-More/Mental-Health-Conditions/Related-Conditions/Suicide.

$201 billion in 2013: https://www.washingtonpost.com/news/to-your-health/wp/2016/05/19/guess-what-medical-condition-is-the-costliest-its-not-heart-disease-cancer-or-diabetes/?utm_term=.bbe1149ca97c.

1. The Rat's Revenge

7   *major regions of the human brain:* http://www.bic.mni.mcgill.ca/Servic esAtlases/ICBM152NLin2009;   https://surfer.nmr.mgh.harvard.edu/ fswiki/FreeSurferMethodsCitation.

13   *three million in the United States:* https://www.nimh.nih.gov/health/sta tistics/prevalence/schizophrenia.shtml.

14   *performed worldwide:* Gordon M. Shepherd, *Creating Modern Neuro-science: The Revolutionary 1950s* (New York: Oxford University Press, 2010).

15   *1993 in* Neuropsychopharmacology: Barbara K. Lipska, George E. Jaskiw, and Daniel R. Weinberger, "Postpubertal Emergence of Hyperrespon-siveness to Stress and to Amphetamine After Neonatal Excitotoxic Hip-pocampal Damage: A Potential Animal Model of Schizophrenia," *Neuro-psychopharmacology* 9 (1993): 67–75, doi:10.1038/npp.1993.44.

developing novel antipsychotic treatments: "Rat or Mouse Exhibiting Behaviors Associated with Human Schizophrenia," U.S. patent no. 5,549,884, issued August 27, 1996, by the United States Patent and Trade-mark Office.

2. The Vanishing Hand

32   *three or more tumors:* https://www.aimatmelanoma.org/stages-of-mel anoma/brain-metastases/.

3. Into My Brain

40   *diagnosed in about 130,000 people each year:* https://www.aimatmela noma.org/about-melanoma/melanoma-stats-facts-and-figures/.

53   *CA209-218:* Expanded Access Program with Nivolumab in Combi-nation with Ipilimumab in Patients with Tumors Unable to Be Re-moved by Surgery or Metastatic Melanoma, ClinicalTrials.gov identi-

fier NCT02186249, https://clinicaltrials.gov/ct2/show/NCT02186249? term=CA209-218&rank=1.

4. Derailed

69 *suddenly went off:* "Phineas Gage: Neuroscience's Most Famous Patient," Smithsonian.com, http://www.smithsonianmag.com/history/ phineas-gage-neurosciences-most-famous-patient-11390067/.

5. Poisoned

81 *midline of the brain:* Michele L. Ries et al., "Anosognosia in Mild Cognitive Impairment: Relationship to Activation of Cortical Midline Structures Involved in Self-Appraisal," *Journal of the International Neuropsychology Society* 13, no. 3 (May 2007): 450–61.
damage to the right hemisphere: Mental Illness Policy, https://mentalill nesspolicy.org/medical/anosognosia-studies.html.
accept their diagnoses: Ibid.
very resistant to psychiatric treatment: Rachel Aviv, "God Knows Where I Am," *New Yorker,* May 30, 2011.
participate in behavioral therapies: C. Arango and X. Amador, "Lessons Learned About Poor Insight," *Schizophrenia Bulletin* 37, no. 1 (January 1, 2011): 27–28.
83 *frontotemporal dementia:* Nadene Dermody et al., "Uncovering the Neural Bases of Cognitive and Affective Empathy Deficits in Alzheimer's Disease and the Behavioral-Variant of Frontotemporal Dementia," *Journal of Alzheimer's Disease* 53, no. 3 (2016): 801–16.
60 to 80 percent of all dementia cases: 2015 Alzheimer's Disease Facts and Figures, Alzheimer's Association, https://www.alz.org/facts/downloads/ facts_figures_2015.pdf.
diagnosed each year: World Health Organization, http://www.who.int/ mediacentre/factsheets/fs362/en/.
forty-five to sixty-four years old: Association for Frontotemporal Degeneration, https://www.theaftd.org/understandingftd/ftd-overview.
84 *typically lack empathy:* Dermody et al., "Uncovering the Neural Bases."
criterion for FTD: K. P. Rankin et al., "Self-Awareness and Personality Change in Dementia," *Journal of Neurology, Neurosurgery, and Psychiatry* 76, no. 5 (2005): 632–39, http://jnnp.bmj.com/content/76/5/632.short.

6. Lost

95 *developmental topographical disorientation (DTD):* G. Iaria et al., "Developmental Topographical Disorientation and Decreased Hippocam-

pal Functional Connectivity," *Hippocampus* 24, no. 11 (November 2014): 1364–74, doi: 10.1002/hipo.22317.

9. What Happened, Miss Simone?

138　*urinary incontinence:* Ryuji Sakakibara et al., "Urinary Function in Elderly People with and Without Leukoaraiosis: Relation to Cognitive and Gait Function," *Journal of Neurology, Neurosurgery, and Psychiatry* 67 (1999): 658–60.

139　*their healthy siblings:* T. M. Hyde et al., "Enuresis as a Premorbid Developmental Marker of Schizophrenia," *Brain* 131 (September 2008): 2489–98, doi: 10.1093/brain/awn167.

140　*"not what happened at all":* T. Rees Shapiro, "Harvard-Stanford Admission Hoax Becomes International Scandal," *Washington Post,* June 19, 2015.

10. The Light Gets In

149　*BRAIN initiative:* https://www.braininitiative.nih.gov/.

11. Survivor

164　*"Neuroscientist Who Lost Her Mind":* Barbara K. Lipska, "The Neuroscientist Who Lost Her Mind," *New York Times,* March 12, 2016, https://www.nytimes.com/2016/03/13/opinion/sunday/the-neuroscientist-who-lost-her-mind.html.

Epilogue

180　*"Soviets and Polio":* Jake Halpern, "A Triathlon Is Easy Next to Soviets and Polio," *Wall Street Journal,* May 22, 2017, https://www.wsj.com/articles/a-triathlon-is-easy-next-to-soviets-and-polio-1495492959.